Half-metallic Materials
and
Their Properties

Series on Materials for Engineering

ISSN:

Editor-in-Chief: William (Bill) E. Lee (*Imperial College London, UK*)

MATERIALS FOR
ENGINEERING Vol. 2

Half-metallic Materials
and
Their Properties

C. Y. Fong (University of California, Davis, USA)

J. E. Pask (Lawrence Livermore National Laboratory, USA)

L. H. Yang (Lawrence Livermore National Laboratory, USA)

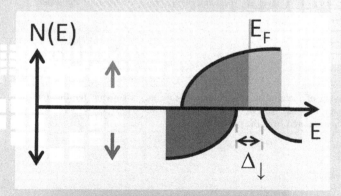

Imperial College Press

ICP

Published by

Imperial College Press
57 Shelton Street
Covent Garden
London WC2H 9HE

Distributed by

World Scientific Publishing Co. Pte. Ltd.
5 Toh Tuck Link, Singapore 596224
USA office: 27 Warren Street, Suite 401-402, Hackensack, NJ 07601
UK office: 57 Shelton Street, Covent Garden, London WC2H 9HE

British Library Cataloguing-in-Publication Data
A catalogue record for this book is available from the British Library.

Materials for Engineering — Vol. 2
HALF-METALLIC MATERIALS AND THEIR PROPERTIES
Copyright © 2013 by Imperial College Press

ISBN 978-1-908977-12-0

Typeset by Stallion Press
Email: enquiries@stallionpress.com

Printed in Singapore by B & Jo Enterprise Pte Ltd

To our families

Preface

Since the first theoretical prediction of a half-metallic material — one in which the electrons of one spin participate in conduction while those of the other do not — was published in 1983, much effort has been exerted to understand, predict, and grow new half-metallic materials. A main impetus for this effort has been the great potential for half-metals (HMs) in spintronics — a new generation of electronics in which the electron spin, as well as its charge, is exploited to achieve substantial reductions in size and/or improvements in performance. In an HM, the electrons of one spin channel are metallic while those of the other are insulating, the spin polarization at Fermi level is complete (100%), and the magnetic moment is an integer. Partially spin-polarized spintronic materials, such as Co/Fe layered structures, have already led to breakthroughs in information storage and processing technologies, such as magnetoresistance (MR)-based read heads, magnetic random access memory (MRAM), and spin-current switches. Because HMs exhibit very large (ideally, infinite) MRs, it is not surprising that the design and realization of HM devices have been among the hottest topics in condensed matter physics and materials science in recent years.

Tremendous theoretical, experimental, and technological progress in the understanding of half-metallic materials has been achieved in the past few years, and the scientific literature has grown correspondingly. It is opportune, therefore, to summarize the main concepts, results, and advances in order to spark and facilitate new research and provide a solid and coherent foundation for new researchers in this exciting field. These are the goals we set out to achieve in the present monograph, at a level appropriate for advanced undergraduate or graduate students in physics, chemistry, and materials science.

In the introductory chapter, we give an overview of the main features, similarities, and differences among the three main classes of HMs which have emerged to date. We discuss key features from both a theoretical and technological point of view. We emphasize the three basic interactions underlying half-metallic properties in all such materials in the hope that new half-metallic materials may be realized by tuning each of the interactions independently. For device applications, it is necessary to understand both electronic and magnetic properties. Substantial progress has been achieved both experimentally and theoretically in the endeavor to understand and predict the properties of half-metallic materials. In Chapter 2, we provide a discussion of the key experimental and theoretical techniques, as appropriate for each class of material, which have been employed to date in order to achieve that understanding and predictive capability.

The three main classes of half-metallic materials that have emerged to date are: (i) the Heusler alloys; (ii) transition-metal oxides; and (iii) pnictides, chalcogenides, and carbides with the zincblende (ZB) structure. Each has its own rapidly evolving literature. We discuss each class in detail in Chapters 3, 4 and 5, respectively. In each chapter, we discuss the crystal structure, experimental characterization, roles of the basic quantum mechanical interactions, and resulting electronic, magnetic, and transport properties.

Quantum structures with tunable properties can be constructed from HMs in various combinations with normal metals, semiconductors, and insulators. In Chapter 5, we discuss some of the more common configurations such as superlattices, quantum dots, digital ferromagnetic heterostructures (DFH), and quantum wires. For superlattices, we include a discussion of active research areas such as transport properties. There is as yet relatively little known about quantum dots in the context of HMs. However, capping and competition of ferromagnetic (FM) and antiferromagnetic (AFM) phases have emerged as key issues. DFHs constitute an ideal configuration to investigate magnetic coupling, and mechanisms to enhance coupling between transition-metal elements, in particular. We include a discussion of theoretical and experimental work in this rapidly evolving area. Quantum wires have emerged as another promising configuration for device applications. In these, the half-metallic properties have been shown to be robust with respect to the spin–orbit interaction and lattice vibrations. We include also a discussion of the similarities and differences of orbital hybridizations in ZB structures and quantum wires.

C. Y. Fong, J. E. Pask, and L. H. Yang

Acknowledgments

We are grateful for the support of the National Science Foundation with Grants ESC-0225007 and ECS-0725902. Work at Lawrence Livermore National Laboratory was performed under the auspices of the U.S. Department of Energy under Contract DE-AC52-07NA27344. We thank Professor Kai Liu for a critical reading of portions relating to experiments, Dr. R. Dumas for providing a diagram of the setup of the radio frequency sputtering scheme, Dr. R. Rudd for a number of helpful comments throughout the writing, and Dr. Michael Shaughnessy for a critical reading of Chapter 5. We also thank the many research groups who have graciously allowed us to include their results.

Contents

Chapter 1

Introduction

1.1. Background

The primary ingredient of computing hardware is the integrated circuit (IC). The number of transistors that can be placed on an IC is characterized by Moore's law (Moore, 1965) which describes a long-term trend in the history of computing hardware. According to Moore's law, the packing density of transistors on a chip doubles approximately every two years. In the last decade, advances in magnetic hard drive technology have increased the capacity of a typical hard drive from tens of gigabytes (GB) to terabytes (TB), while the bit density of magnetic heads has increased 60–100% annually to 250 gigabits/in^2, which suggests the law still holds. However, there is a limit to such packing due to the finite interatomic distance of a few Å in typical materials,[1] suggesting that Moore's law must eventually be violated. Furthermore, another important issue is the volatility of the information stored. For example, in dynamic random access memory (DRAM), data is stored in the form of charge on capacitors. Therefore, whenever the power is switched off the information is lost. The approaching limit in packing density and volatility of information stored must be overcome to produce smaller, faster devices.

Innovative technologies have been developed since the discovery of giant magnetoresistance (GMR) (Baibich *et al.*, 1988; Wolf *et al.*, 2001) to utilize the electron spin degree of freedom for information manipulation, storage, and transmission. Metal-based devices in the form of heterostructures have already found applications in ultra-high density magnetic recording (Ross, 2001), as GMR sensors (Baibich *et al.*, 1988; Parkin *et al.*, 1991; Dieny *et al.*, 1992), in prototype magnetic random access memory (MRAM) (Daughton, 1992; Bussmann *et al.*, 1999; Savtchenko *et al.*, 2003), and in spin-current switches (Albert *et al.*, 2000; Grollier *et al.*, 2001). These

---

[1] http://www.almaden.ibm.com/st/

devices utilize ordinary ferromagnetic (FM) materials, such as ion (Fe), cobalt (Co), and nickel (Ni), that have a spin imbalance, or finite electronic spin polarization P at Fermi energy E_F. Information can be encoded into the finite resistance change resulting from spin-dependent electric transport through these magnetic heterostructures. The advantages of these devices over conventional nonmagnetic devices are their increased integration densities, nonvolatility, faster speed in data processing, and lower power consumption.

To develop further these new technologies exploiting electron spin, it is advantageous to leverage established technologies. One emerging field combines semiconductors and magnetism. The term "spintronics" was coined by S. Wolf in 1996 (Wolf *et al.*, 2001). It is characterized by the transport of spin, or charge and spin, in semiconductor-based materials. Therefore, a crucial initial step in developing next-generation spin-based technologies is the development of spintronic materials. These materials can be composed of FM metals, magnetic semiconductors with or without doping, and/or so-called "half-metals" (HMs). Such materials stand to enable new technologies with dramatically increased speeds, decreased sizes, and nonvolatile storage. Figure 1.1 illustrates the development of these next-generation technologies. Materials issues are crucial concerning design, growth, and contact to electrodes to efficiently transfer spin. Associated devices may be expected to include spin field-effect transistors (FETs), memory devices, and components for quantum computing.

Normally, a nonferromagnetic material, such as silicon (Si) for example, has spin degeneracy in its energy states — i.e., energies are independent of spin and occupation of spin-up ($\uparrow$) and spin-down ($\downarrow$) states is equal,

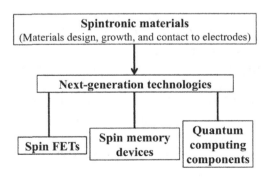

Fig. 1.1.   Developing next-generation spin-based technologies and applications.

leading to zero net spin polarization. In an FM metal, such as Fe, on the other hand, this degeneracy is broken and more states of one spin channel (the "majority" channel) are occupied than the other ("minority" channel), leading to nonzero net spin polarization and FM properties. Since coordinate axes can be chosen as desired, it is typical to denote, without loss of generality, the majority-spin channel as ↑ and the minority-spin channel as ↓ and we shall adopt this convention here. The spin polarization P_N is defined by

$$P_N = \frac{N_\uparrow - N_\downarrow}{N_\uparrow + N_\downarrow},$$ (1.1)

where $N_\uparrow$ and $N_\downarrow$ are the number of ↑ spin and ↓ spin states at E_F, respectively. Under this definition, P_N measures the spin imbalance of mobile electrons. A typical value for P_N in FM metals is 40~50% at room temperature (RT). It has also been defined, alternatively, as the net fractional spin polarization near E_F. In this case, we denote it by P to distinguish from P_N, with

$$P = \frac{d_\uparrow - d_\downarrow}{d_\uparrow + d_\downarrow},$$ (1.2)

where $d_\uparrow$ and $d_\downarrow$ are the density of states (DOS) of ↑ spin and ↓ spin channels at E_F, respectively. P can therefore be directly determined from the DOS.

de Groot *et al.* (1983) predicted a remarkable property in the half-Heusler alloy, NiMnSb. In this compound, the ↑ spin states exhibit metallic properties while the ↓ spin states are insulating. At E_F, $N_\downarrow$ and $d_\downarrow$ vanish identically. Consequently, all states at E_F are ↑ spin and NiMnSb has $P = 100\%$. These properties define "half-metallicity" and materials possessing these properties are called "half-metals". The earliest Heusler alloys were grown by Heusler (1903). Materials such as NiMnSb are called half-Heusler alloys. Many of the so-called full-Heusler alloys, e.g., Co_2MnSi, are also predicted to be HM theoretically, and some have been confirmed experimentally. The difference between half- and full-Heusler alloys lies in the occupation of sites within the unit cell, as we discuss further below.

There are also predictions of HMs in other structures, such as oxides (Schwarz, 1986) in the rutile structure. Among them, only CrO_2 has been verified experimentally as an HM at low temperature. Other oxides as well as chromium arsenide (CrAs) (Akinaga *et al.*, 2000a) and manganese carbide (MnC) (Pask *et al.*, 2003) in the zincblende (ZB) structure were predicted theoretically to be HMs. In the latter compound, the usual roles played by the majority- and minority-spin channels are reversed.

In all of the above cases, the transport properties are determined solely by electrons in states in the vicinity of E_F with a single spin polarization. In some applications, both charge and spin transport can be envisioned. Transport utilizing the spin degree of freedom provides completely new prospects for information storage and transmission. It is, all but certainly, only a matter of time before this type of transport, with all its possibilities, is incorporated into semiconductor technologies. However, there are concerns about the disappearance of half-metallicity at RT in these materials. Those materials experimentally demonstrated to reach over 95% spin polarization, such as CrO_2, cannot sustain their spin polarization above RT (Dowben and Skomski, 2004). Structural transitions, collective excitations (e.g., spin waves and phonons), correlations associated with onsite Coulomb interactions, and spin polarons (Katsnelson *et al.*, 2008) which show non-Fermi liquid behavior and are formed in the gap near E_F, can cause loss of half-metallicity. In this monograph, we focus mainly on the design, growth, and basic understanding of the electronic and magnetic properties of half-metallic materials determined experimentally at low temperature and predicted theoretically at or near $T = 0$ K. We will, however, briefly comment on issues at higher temperature as appropriate.

1.2. Classes of Half-metals

Up to now, three main classes of FM HMs have been found by theory and/or experiment with distinct crystal structures. They are the Heusler alloys (de Groot *et al.*, 1983; Galanakis, 2002b), such as Co_2CrAl; the oxides, such as CrO_2 (Schwarz, 1986); and those having the ZB structure (e.g., CrAs) (Akinaga *et al.*, 2000a; Pask *et al.*, 2003) or diamond structure (Qian *et al.*, 2006a). The unit cell of a typical full-Heusler alloy is shown in Fig. 1.2 and that of CrO_2 is shown in Fig. 1.3. In Fig. 1.2, the Co atoms are shown as filled and open circles, the Cr atoms are denoted by open triangles, and the Al atoms, by open squares. A common feature of all these HMs is that they contain at least one transition-metal (TM) atom, such as Co, Mn, or Fe, in the unit cell. Their d-states play a key role in the half-metallicity. The interaction of the d-states with the states of other atoms — e.g., the d-states of other TM atoms, the p-states of oxygen atoms, pnictides, or Group IV elements — gives rise to a number of distinguishing properties.

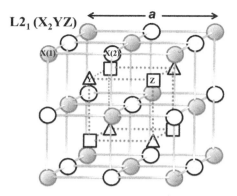

Fig. 1.2. L2$_1$ crystal structure of full-Heusler alloy (X$_2$YZ), such as Co$_2$CrAl. The X=Co atoms are denoted by filled and open circles, the Y=Cr atoms by open triangles, and the Z=Al atoms by open squares. The outermost cube edge has length a. If the sites with open circles (X(2)) are unoccupied, the structure corresponds to a half-Heusler alloy (XYZ) and is denoted by C1$_b$ (Galanakis *et al.*, 2002a).

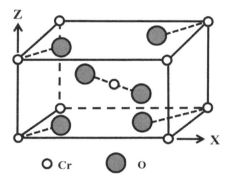

Fig. 1.3. Crystal structure of CrO$_2$.

To achieve a basic understanding of the half-metallic properties and differences among the three classes, we give an overview in terms of the DOS based on one-electron properties. The DOS for each class shows distinct features as a consequence of the atoms involved and distinct crystal structure. A typical DOS exhibits s-like states from the non-TM atoms, such as oxygens, pnictides, and Group IV elements, in the low-energy region of the valence manifold. The d-states of the TM are split into triply and doubly degenerate multiplets in a cubic or tetrahedral environment. In the cubic case, the triply degenerate states are labeled t$_{2g}$ and have lower energy than the doubly degenerate states labeled e_g because the lobes of the d_{xy}, d_{yz}, and d_{zx} states point toward neighboring atoms while the lobes of the e_g states point toward second-nearest neighbors.[2] In the tetrahedral environment, the order of the two sets is reversed. Depending on the strength of the d–d interaction between neighboring TM elements relative to the d–p interaction between TM and neighboring non-TM elements, the highest occupied states can be either d or d-p hybrid in nature. The FM exchange interaction determines the relative energies of the majority- and minority-spin states and the value of E_F. Considering the d–d and d–p interactions, the half-metallicity in all three classes of compounds can be understood in terms of the crystal field, hybridization, and exchange interaction.

In Fig. 1.4, we show the DOS near E_F of a half-Heusler alloy. Typically, a Heusler alloy is metallic in the majority-spin channel. In this case, the d–d interaction between the X and Y atoms (Galanakis *et al.*, 2002a) determines the states near E_F. The Fermi level intersects the e_g states in the majority-spin channel. There is an overlap between the e_g and t_{2g} states in this channel due to the d–d interaction. In the minority-spin channel, the gap is formed between the hybridizing d (X-atom) and p (Z atom)

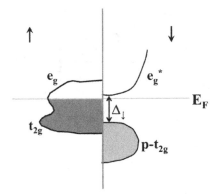

Fig. 1.4.   Schematic DOS of a half-Heusler alloy. $\Delta_\downarrow$ is the energy gap for the $\downarrow$ spin states.

---

[2]The t_{2g} and e_g labels are not the only ones used to denote triply and doubly degenerate states, especially in condensed matter physics. For example, the labels $\Gamma_{25'}$ and $\Gamma_{12'}$ are also used in cubic cases (having inversion symmetry). In the ZB structure, Γ_{15} is used to label the three-fold degenerate states. In recent years, however, the t_{2g}/e_g notation has become more prevalent (even when not strictly applicable to the symmetry at hand) and we shall use this convention here.

states (p-t$_{2g}$ hybrid) and antibonding d states (e$_g^*$) from the TM (Y-atom). A common feature of the Heusler alloys is that E_F is located just below the unoccupied $\downarrow$ spin states — i.e., bottom of the conduction band in the minority-spin channel. This feature has important consequences, as we describe later.

Coey and Venkatesan (2002) classified the DOS of half-metallic oxides into three types. They are I$_A$, I$_B$, and II$_B$ and are shown in Figs. 1.5, 1.6, and 1.7. The label A indicates that the $\uparrow$ spin channel is conducting. The label B indicates that the $\downarrow$ spin channel is conducting. In all three types, the lowest-energy s-states are below the energy range shown. The next group of states are oxygen 2p in nature. They are bonding states.

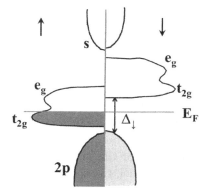

Fig. 1.5. Type I$_A$ DOS, $\Delta_\downarrow$ is the insulating gap for the $\downarrow$ spin channel.

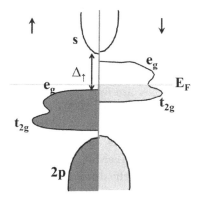

Fig. 1.6. Type I$_B$ DOS.

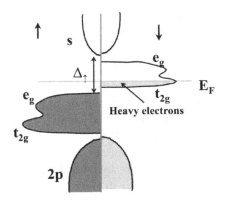

Fig. 1.7.   Type II$_B$ DOS.

Both ↑ spin and ↓ spin channels are equally occupied. They experience a small effect from FM exchange.

For the type I$_A$ HMs (Fig. 1.5), the next structure in the DOS is derived from the d-states of the majority-spin channel. They are partially occupied and the Fermi energy E_F passes through the t_{2g} states. The e_g states overlap with the t_{2g} states and are located at higher energy. This is a consequence of the octahedral crystal field. The next higher energy structure is the unoccupied d-states of the minority-spin channel. Just as for the occupied majority-spin channel, the e_g states are at higher energy and overlap with the t_{2g} states. The insulating (semiconducting) gap ($\Delta_\downarrow$) is in the minority-spin channel between the t_{2g} states and bonding oxygen p-states. An example of this type of HM is CrO_2. As shown later, the difference between this schematic DOS and the calculated DOS is that the latter shows some d-p hybridization in the occupied states, with dominant oxygen p.

For the type I$_B$ HMs (Fig. 1.6), the next structure above the bonding oxygen p-states is also derived from the d-states of the majority-spin channel. In this case, however, they are fully occupied. The next higher energy structure is the overlapping t_{2g} and e_g states of the minority-spin channel with E_F passing through t_{2g} states. The insulating gap is formed between fully occupied majority-spin d-states and anti-bonding oxygen s-states. An example of this type of HM is Sr_2FeMnO_6.

For type II$_B$ HMs (Fig. 1.7), the d-manifolds of majority- and minority-spin states do not overlap. The lower energy majority-spin states are fully occupied. The mechanism of conduction differs from that of type I$_A$ and

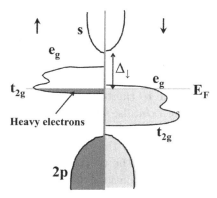

Fig. 1.8. Type III$_A$ DOS.

I$_B$ HMs. The minority t$_{2g}$ electrons form polarons. Fe$_3$O$_4$ is an example of this type of compound. The half-metallic properties of this type of HM, like others, can be affected by impurities. For example, if the sublattice occupied by TM elements is substituted by other elements, the sample can become a Mott insulator.

Based on the mobility μ and the effective mass m^* of the mobile carriers, rather than integer magnetic moment/unit cell, Coey and Venkatesan (2002) considered two additional types of HM (types III$_A$ and IV$_A$). As shown in the DOS (Fig. 1.8) for a type III$_A$ HM, both spin channels intersect E_F. However, the majority-spin electrons are localized while the minority-spin electrons are delocalized. There is thus a large difference of μ and m^* for carriers of different spin. The electrons in the majority-spin channel are essentially immobile; and so conduction is confined mainly to the minority-spin channel.

The main features of type I and II HMs can be qualitatively understood in terms of two key facts: (i) oxygen atoms have large electronegativity and (ii) TM elements have tightly bound d-states. Consequently, the oxygen atom essentially ionizes the electrons of neighboring cations to fill its 2p-states. The d-states of the cation split into three-fold t$_{2g}$ and two-fold e$_g$ states in the octahedral field of the surrounding oxygen atoms. The exchange interaction shifts the energies of the minority-spin states up relative to the majority-spin states. The occupied d-states remain localized at the cation site with more or less atomic-like features. The p-states of the oxygen atoms show less hybridization with the d-states than do the p-states of the pnictide or Group IV elements in the Heusler alloys due to the large electronegativity of the oxygen atoms. There is no d-d hybridization because the TM elements are surrounded by the oxygen atoms.

For an HM with the ZB structure, we first give a qualitative discussion of the bonding and then comment on key features of the DOS. The anion can be a valence IV, V, or VI element. Its electronegativity is in general weaker than that of an O atom. Due to the tetrahedral environment, the anion s and p states form sp^3 type orbitals, which point toward the neighboring cations. The cation, a TM element, has its five-fold degenerate d-orbitals split into t_{2g} and e_g-type states in the tetrahedral environment. The t_{2g} states are higher in energy than the e_g states and are comprised of d_{xy}, d_{yz}, and d_{zx} states. Linear combinations of these form directional orbitals pointing toward neighboring anions. These orbitals can interact with the sp^3 type orbitals of the anions to form bonding and antibonding states. Figure 1.9 shows the structure and d-p hybridization in MnAs. On the left-hand side, the As atoms are indicated by filled circles and the Mn atoms by open circles. One Mn atom is located at $(1/4,1/4,1/4)a$ along the cubic body diagonal, where a is the length of the cube edge. The primitive cell is defined by one Mn atom and one As atom. On the right-hand side of Fig. 1.9, the overlap of an Mn d-orbital and As sp^3 orbital is depicted. This overlap gives rise to bonding and antibonding states. The bonding states have covalent character, i.e., charge sharing with neighbors. These are the d-p hybrid states. The energy of the bonding p-t_{2g} hybrid states is lower than the e_g states. The ordering of energies is shown schematically in Fig. 1.10. At the left and right ends, the energy levels of the d-states of a TM element and the s- and p-states of a chalcogenide, pnictide, or

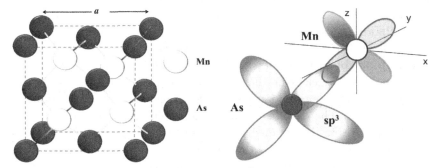

Fig. 1.9. MnAs: As atoms (filled circles) are at the corners and face centers. An Mn atom (open circle) is surrounded by four As atoms and is located at $(1/4,1/4,1/4)$ along the body diagonal. A schematic diagram of the d-p hybridization is shown on the right-hand side.

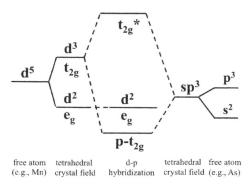

Fig. 1.10. Schematic diagram of the crystal-field splitting of 3d orbitals for one spin under tetrahedral symmetry and d-p hybridization in the ZB structure. Superscripts indicate degeneracy.

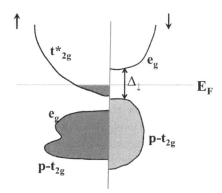

Fig. 1.11. Schematic DOS of an HM with the ZB structure.

carbide are shown. Moving toward the center, the effects of the crystal field are illustrated. For the TM atom, the five-fold degenerate d-states split into triply degenerate t_{2g} and doubly degenerate e_g states. The non-TM element forms sp^3-type orbitals. At the middle of the figure are the bonding (p-t_{2g}) and antibonding (t_{2g}^*) states resulting from the d-p hybridization.

The ordering of these states is shown in the schematic DOS of Fig. 1.11. Only states in the vicinity of E_F are shown. The e_g states are comprised of d_{z^2} and $d_{x^2-y^2}$ states. They point toward second neighbors rather than nearest neighboring cations and form the nonbonding states or bands in the crystal. They can overlap in energy with the d-p bonding states (p-t_{2g}) or can be separated from them to form a gap.

As shown in the ↓ spin channel, a gap exists between the bonding p-t_{2g} states and nonbonding e_g states. E_F passes through this gap. For the majority-spin states, an overlap between the bonding p-t_{2g} and nonbonding e_g states is typical, as shown in the figure. The bonding states have significant anion-p character. To accommodate the total number of valence electrons in the unit cell, the lowest-energy antibonding states in the majority-spin channel are occupied due to the exchange splitting of majority- and minority-spin states. These antibonding states are d-p hybrid states with predominantly transition-element d character. This partial occupation in the majority-spin channel gives rise to the half-metallicity.

Among the three classes of HMs, the d-states are dominant near E_F in the Heusler alloys and the oxides. Hybridization is the strongest in the ZB HMs. The states at E_F for these are d-p hybrid in character.

1.3. Half-metallic Devices

Since HMs offer the possibility of unprecedented magnetoresistance (MR) (infinite in principle), there is huge potential for device applications, improving substantially the majority of GMR-based devices in current use and offering a host of new possibilities. Hence, this has been an active field of research. There have been many theoretical proposals, based on extensive quantum mechanical calculations, as we detail in subsequent chapters. Here, we note some recent progress on practical device realization.

Recent efforts in spintronic device fabrication have focused mainly in the areas of spin valves and spin transistors. The latter was proposed by Datta and Das (1990) and is similar to standard FETs. However, the requirements for effective FETs, such as large polarized current, capability of amplification, and simple material structure, are extremely demanding, and there has not yet been successful fabrication of practical devices in this form. There has been more progress in fabricating HM-based spin valves. In these devices, spin is injected from HMs to semiconductors. It has been identified by Wang and Vaedeny (2009) that four conditions should be met for an effective spintronic device. Among the four, the most important is efficient spin injection from the metallic materials normally serving as electrodes into semiconductors. To maintain the signal, electrons in the semiconductor should have long lifetimes for conduction and spin relaxation. In crystalline semiconductors, the existence of the spin–orbit interaction can change the spin moment of the carriers. Wang and Vaedeny used organic semiconductors instead of conventional semiconductors. They made a device with $La_{2/3}Sr_{1/3}MnO_3$ (LSMO) and Co serving as electrodes. The organic semiconductor 4,4′–bis–(ethyl–3–carbazovinylene)–1,1′–biphenyl (CVB) was used to reduce the effect of the spin–orbit interaction. The measured spin-polarized carrier density was 60–20% in the temperature range of 60–120 K. Kodama *et al.* (2009) grew the layered structure $Cr/Ag/Cr/$ $Co_2MnSi/Cu/Co_2MnSi/Co_{0.75}Fe_{0.25}/Ir_{0.22}Mn_{0.78}/Ru$ on MgO (001) substrate. Cu was included between Heusler layers to facilitate the transport of spin-polarized carriers from one Heusler alloy to the other, which would be impeded by Cr. The Ag layer between the Cr layers served to enhance the perpendicular current. The ratio of MR was found to be 8.6% at RT and 30.7% at $T = 6$ K. Hence, more work is needed to get MR and robustness to desired levels, and much work continues.

Chapter 2

Methods of Studying Half-metals

2.1. Introduction

To realize spintronic devices based on HMs, the growth of these materials must be addressed. Growth methods for the three classes of HMs are not universal; some are specific to a particular class or classes. In this chapter, we discuss a method which has been used to grow Heusler alloys and HMs with the ZB structure. Having discussed growth, we next consider methods used to characterize the quality of the samples grown — a crucial step for all subsequent understanding and application. We then consider techniques used to measure the properties of the resulting materials, including electronic properties, magnetic properties, half-metallicity, transport properties, and Curie temperature T_C. Most of the characterization methods and measurement techniques are commonly applied to all classes of HMs. Since de Groot *et al.* (1983) first predicted half-metallic properties in NiMnSb, theoretical modeling has been shown to be extraordinarily effective in understanding and predicting new HMs. This requires methods with substantial predictive power for atomic arrangements quite unlike any known or considered heretofore. Therefore, first-principles methods based on density functional theory (DFT) are among the most popular for this purpose and we include a discussion of the basic concepts and practical implementations of these.

In this chapter, we first discuss the key experimental method used for growing the Heusler alloys and HMs with ZB structure in thin-film form. We then discuss the methods of characterization and techniques for measurement commonly employed. Finally, we conclude with a brief discussion of theoretical methods most commonly employed, since many excellent reviews are available (e.g., Martin, 2004).

2.2. Molecular Beam Epitaxy (MBE)

Because HM crystal structures are not always the stable bulk ground state structures, and because desired MR properties are often achieved by multilayer structures, the ability to grow HM structures in thin-film form is of crucial importance. The molecular beam epitaxy (MBE) method is an ideal way to grow quality thin films. It was originally designed to grow layered semiconductors and semiconductor heterostructures (Chang and Ploog, 1985).

Its application to grow NiMnSb was carried out by Van Roy *et al.* (2000) and Turban *et al.* (2002). Van Roy *et al.* used the Riber 32P chamber while a commercial chamber was used by Turban *et al.* Ambrose *et al.* (2000) also used a commercial chamber to grow Co$_2$MnGe. With the MBE method, Akinaga *et al.* (2000a) were the first group to successfully grow CrAs thin film with ZB structure and Zhao *et al.* (2001) grew ZB CrSb with a single monolayer. Hereinafter, we will refer to these types of samples as HMs with the ZB structure.

2.2.1. *Schematic setup*

A schematic diagram for the setup of the MBE method is shown in Fig. 2.1. The growth chamber is shown as a circle. The residual pressure in the chamber is commonly maintained at the level of 10^{-9} mbar.

The beam sources are indicated by open rectangles connected to the chamber. Depending on the desired number of constituent atoms, more than two beam sources are possible. The sample holder is shown in black. The substrate is on top of the sample holder; and the sample, on top

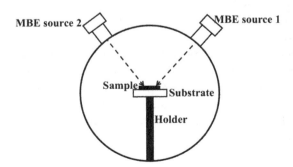

Fig. 2.1.   A schematic diagram of the setup for the MBE method. The substrate is just above the holder. Sample is shown in black horizontal rectangle.

of the substrate. In general, characterizations of the sample are made *in situ* and measurements are carried out without moving the sample out of the chamber. In that case, the chamber will have to accommodate such equipment.

The source cells are, of course, specifically dependent on what is to be grown. For the growth of NiMnSb, a nonvalved cracker cell for Sb is sometimes used. The cracker zone was kept at 930°C. Ni and Mn atoms with 1:1 ratio are furnished (Ambrose *et al.*, 2000). Alternatively, Turban *et al.* (2002) prepared 99.99% pure Ni and Mn and 99.999% pure Sb Knudsen cells. They also used a quartz-crystal microbalance installed at the sample position to monitor the flux of the atomic beams. To grow Co_2MnGe, atomic fluxes are supplied by individual Knudsen cell sources monitored by a quadrupole mass analyzer. For the growth of thin-film forms of CrAs having ZB structure, GaAs (001) serves as the substrate (Akinaga *et al.*, 2000a). Below it, a GaAs buffer layer of the order of 20 nm thick was grown first at 580°C. This helps to obtain a flat surface for the substrate, to reduce the stress of the substrate. The substrate should be annealed at 600°C for about 10 minutes in the chamber in order to remove oxidation layers or other contaminations. During the growth, the substrate is under rotation and is irradiated by Cr and As beams at once. The rate of growth was 0.017 nm/s. After the growth, the sample was capped by low temperature GaAs to prevent oxidation when the sample is exposed to air. The thickness of the capping layer was 5 nm.

2.2.2. *Issues concerning growth*

MBE has been applied to the growth of Heusler alloys and HMs with ZB structure. The quality of a thin-film sample depends on the selection of the substrate and the control of the substrate temperature. These are general issues for all thin-film growths but they are sample-specific. The following discussions on these two issues will focus on Heusler alloys and HMs with ZB structure.

2.2.2.1. *Substrate for Heusler alloys*

For Heusler alloys, both growths by Van Roy *et al.* (2000) and Ambrose *et al.* (2000) used GaAs(001) as the substrate based on the compatibility of the crystal structures between the growth sample and substrate and based on the matching of lattice constants between them. Additionally, Turban *et al.* (2002) used single-crystal MgO with (001) surface. MgO has a relatively smaller lattice constant which is 4.21 Å. There should be a large lattice-constant mismatch. This kind of substrate, in general, can be used to grow polycrystalline samples in film forms (Bauer, 2010). Therefore, as described above, the primary consideration for the choice of a substrate is determined by matching lattice constants between the sample and the substrate. Alternative choices to grow single-crystal thin films can be made based on experience by adding additional buffer layers.

2.2.2.2. *Substrate for HMs with ZB structure*

For growths of HMs with ZB structure, it is more critical to take the matching of lattice constants between the thin film and substrate into consideration because the ZB structure for these compounds is not the ground state structure. If there is additional stress due to lattice-constant mismatch, the metastable ZB HM will cause devices made of these materials to have limited lifetimes. Another possibility is that the half-metallicity can be lost or obtained under stress. As we discussed earlier, Akinaga *et al.* (2000a) used GaAs(001) as the substrate for CrAs growth, while Zhao *et al.* (2001) grew CrSb on GaSb.

2.2.2.3. *Temperature*

The temperature of the substrate is an important factor to determine the quality of the films. For growing NiMnSb on GaAs(001), the best films are obtained when the temperature of the substrate T_{sub} is kept at 300°C (Van Roy *et al.*, 2000). Turban *et al.* (2002) demonstrate clearly that T_{sub} at about 350°C is the best to grow NiMnSb by measuring the intensity of the reflection high-energy electron diffraction (RHEED) to be described later. On the other hand, the substrate temperature is maintained at 175°C for Co_2MnGe growths. To grow a better CrAs, the temperature of the substrate cannot be maintained at 580°C required to have a flat surface of the substrate. It is necessary to bring down the temperature to 200°C and 300°C range under As pressure.

During or after a sample is grown, it is necessary to characterize its quality and to measure its physical properties. In the following two sections, we will discuss methods of characterization, in particular samples in thin-film forms, and of determining various physical properties relevant to spintronic applications. These methods are generally applied to all HMs. The results, however, will be sample-dependent and discussed specifically for each sample.

2.3. Characterization of Samples

Two popular methods used to characterize the quality of HMs grown in thin-film form are RHEED (Harris *et al.*, 1981) and low-energy electron diffraction (LEED) (Heinz, 1995). The thickness of a film can be determined by either high-resolution X-ray diffraction (HRXRD) (Mukhamedzhanov *et al.*, 2000) or X-ray reflectometry (XRR) (Toney and Brennan, 1989). To correlate whether one period of oscillation corresponds to the growth of one atomic layer or of one unit cell depends on the sample. Methods to distinguish such correlations are XRR and scanning tunneling microscopy (STM) (Binnig *et al.*, 1983).

Although RHEED is a powerful method, it is noted that the results of the method cannot distinguish between chemically ordered and disordered structures; in particular, the degree of intermixing between atoms of the sample and substrate. In this respect, Auger electron spectroscopy (AES) (Wang *et al.*, 2005b) is an appropriate method and should be carried out. Between the RHEED and LEED methods, RHEED method provides more information. Since it is the more powerful method of the two, we shall give a detailed discussion. This will be followed by methods of XRR and STM. The former can serve two functions as described above and the latter is now commonly used. Since intermixing near the layer boundary is one of the important issues relevant to the quality of the film, we shall discuss AES afterward.

2.3.1. *Reflection high-energy electron diffraction (RHEED)*

2.3.1.1. *Basic information*

A beam of electrons is used as the probe in this method. The energy of the electron is in the range of 15.0 keV. The beam is incident at a glancing angle of about 2.0° on the sample. The reflected electron beam provides the interesting information.

2.3.1.2. *What is measured*

There are two quantities to be measured: the interference pattern of the reflected beam that is called the RHEED pattern and the intensity at the center labeled as (00) of the pattern.

RHEED pattern   The pattern is in striped form and provides the structural information of the sample. The primary stripes are the constructive

$[1\bar{1}0]_{GaAs}$

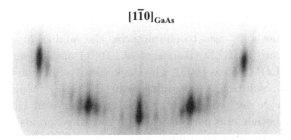

Fig. 2.2.   A typical RHEED pattern of a surface with streaks (Wang *et al.*, 2005b). The primary ones are shown in dark regions. The streaks are shown in light stripes.

interference pattern relevant to the fundamental structure of the sample surface. Therefore, one of the features of RHEED is the capability to characterize a surface structure. Before the growth, it is important to check the pattern of the substrate. When the primary stripes coincide with those of the substrate, lattice-constant matched growth can be inferred. If there are additional weak streaks, it indicates the surface is reconstructed. A typical RHEED pattern for a surface structure is shown in Fig. 2.2.

The pattern shown in Fig. 2.2 is for a clean GaAs with electron beam focused along the $[1\bar{1}0]$ crystal axis. The stripes with elongated dark regions are the primary lines. Spacing between these primary lines can fill with other weak streaks which are related to the reciprocal lattice vector in the crystallographic direction of the surface for the sample with a reconstruction. This is one way to extract lattice-constant information of the sample. In this figure, there is clearly a reconstruction with the periodicity of six times the lattice constant — each primary line is accompanied by five weak streaks.

Oscillatory behavior of spot intensity at the center (00)   This oscillatory behavior reveals the growth mode is in the layer-by-layer form or the Frank–van der Merwe growth mode (see Bauer, 2010). The absence of this behavior indicates the growth is not in the layer-by-layer mode. It is not observed at the beginning of the growth process since there are no complete layers. The general feature of this intensity is an initial drop. Then, the oscillatory behavior appears. The reason for the initial drop can be attributed to the intermixing between the growing layer and substrate or poor layer-by-layer growth. After that, the layer-by-layer growth prevails. An example of intensity as a function of time for growing Co_2MnSi is given in Fig. 2.3.

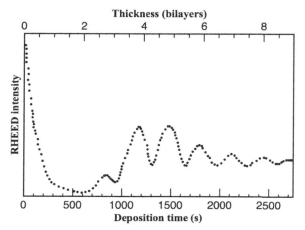

Fig. 2.3.   The oscillatory behavior of RHEED intensity at (00) varying with time (Wang *et al.*, 2005b).

2.3.2. *X-ray reflectometry (XRR)*

Many methods, such as atomic force microscopy (AFM) and X-ray photo-electron spectroscopy are limited by their sensitivities to a few nm. X-ray reflectometry — which was developed for monitoring the quality of transistors, in particular the interface of Si/SiO_2 and the layer thickness of SiO_2 — is still a viable tool for determining the thickness of a film on a nm scale, with an advantage of nondestructivity (Wang *et al.*, 2005b).

2.3.2.1. *Basic information*

The source of X-ray is typically Cu-K lines with wavelength at 0.154 nm for semiconductor applications. Synchrotron radiations with wavelength of the order of 1.0 nm have also been used.

2.3.2.2. *Schematic setup*

A schematic diagram for X-ray reflectometry is shown in Fig. 2.4. The X-ray source is indicated by the light gray rectangle. The X-ray is reflected from a graded multilayer mirror called the Gobel mirror (tilted black rectangle) and is passed to the first slit, S_1, before arriving at the sample. The reflected X-rays pass two slits, S_2 and S_3 to block the stray X-rays, then reach the detector.

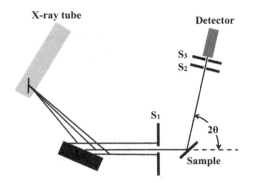

Fig. 2.4.   A schematic setup for X-ray reflectometry.

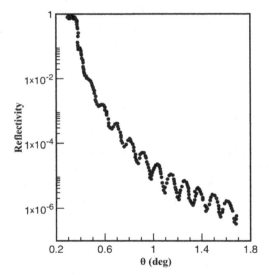

Fig. 2.5.   Measured intensity of specular X-ray reflection as a function of θ (Toney and Brennan, 1989).

2.3.2.3. *What is measured*

Two quantities are measured: the specular reflected X-ray and the diffuse X-ray. They are functions of angle 2θ — the angle between the incident and reflected X-rays. Typical results of these two X-rays are shown in Figs. 2.5 and 2.6. In Fig. 2.5, the specular X-ray signal is measured in units of counting rate (cps) and is shown as a function of θ. At small θ, the intensity is constant. Then, it follows a region of having a negative slope. The angle θ at which the slope changes from zero to negative can be used to determine

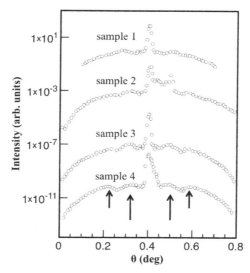

Fig. 2.6. Intensity of diffuse X-ray reflection as a function of θ (Bahr *et al.*, 1993). The Yoneda peaks are marked by arrows.

the density of the substrate. The slope is affected by the roughness of the substrate, the interface roughness and thickness of the film causing the reduction of intensity. The modulations (wavy form) are due to interference of reflections between the surface of the sample and the interface of the sample and substrate. In Fig. 2.6, a typical diffuse signal is plotted by Bahr *et al.* (1993). This set of curves can only be measured when there is surface or interface roughness. The two side peaks come from total external reflection (sample-air) with incident and exit angles at the critical angle of total reflection. They are called "Yoneda wings" (Yoneda, 1963). The center peak originates from specular reflection. The curve without the center peak is determined by the surface layer. A recent development of X-ray reflectometry is reported by Sacchi *et al.* (2007). They used pinholes to control the coherence of the X-ray and constructed a holographic image of the surface.

2.3.3. *Scanning tunneling microscopy (STM)*

2.3.3.1. *Basic information*

STM was developed by Binnig *et al.* (1983). It is an extremely useful tool for surface science. It involves a metallic tip and the sample for which the

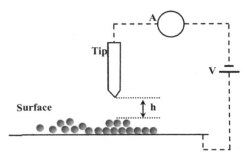

Fig. 2.7. A schematic diagram of STM. h is the separation between the tip and surface. An ammeter (A) is shown as a large circle. The voltage "V" is maintained between the tip and surface.

surface structure is to be determined. Over the years, the technique has been much refined. A schematic diagram is shown in Fig. 2.7. The tip scans across a section of the surface. The height h is the separation between the tip and top layer of the surface. It is controlled by a feedback mechanism to either maintain at an absolute height — to form the STM image providing the morphology of the surface — or at a fixed value with respect to the top layer of the surface, by keeping the measured current constant.

2.3.3.2. *What is measured*

The tunneling current as a function of h at a series of points on the surface is measured. The theory of the tunneling current was worked out by Tersoff and Hamann (1985). Two quantities are obtained from measurements of the tunneling current: if h is kept constant, the current in the ammeter is measured. On the other hand, if a constant current is maintained, then h is measured. Then from the amplitude of the current, an STM image can be deduced. This latter scheme was adopted by Wang *et al.* (2005b). The STM image for Co_2MnSi obtained by Wang *et al.* is shown in Fig. 2.8. By guiding the tip to move along a line, h as a function of position along the line determines the profile of the surface morphology and the thickness of the layer growth. From the height of the profile, it is possible to determine the layer thickness. As shown in Fig. 2.8, the height of $a/2$ (2.85 Å) is obtained, where a is the cubic edge. It reflects the height of islands formed above the substrate during the growth. This distance is half of the cubic edge for the full Heusler alloy.

This STM height determination inferred by the RHEED period of oscillations reflects the growth of two atomic layers instead of four for a full unit

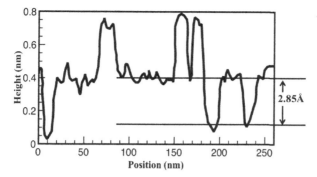

Fig. 2.8. STM line profile along a line parallel to the surface. $a/2 = 2.85$ Å, where a is the lattice constant (Wang *et al.*, 2005b).

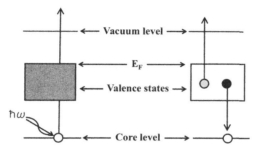

Fig. 2.9. Processes in AES. Left figure shows a core hole (open circle) being created by an incident X-ray. Right figure describes the hole being annihilated by a valence electron (dark circle) and another valence electron (gray circle) being excited above the vacuum level.

cell. This half of the unit-cell growth is required to satisfy the chemical composition and electrical neutrality.

2.3.4. *Auger electron spectroscopy (AES)*

2.3.4.1. *Basic processes*

A photon with energy higher than 1.0 keV excites a core state of an atom in a solid. The hole created in the core region is annihilated by a valence electron, and the energy released from the recombination excites another valence electron. The processes are shown in Fig. 2.9, where the left figure shows a core hole (open circle) being created by the incident X-ray. The right figure describes the hole being annihilated by a valence electron (dark circle) and another valence electron (gray circle) being excited above the vacuum level. The processes involve core states which can exhibit the characteristics of each atom.

2.3.4.2. *Simplified experimental setup*

In most cases, it is desirable to determine the extent of the intermixing of atoms in the substrate *in situ*. Therefore, the setup is more or less identical to the growth process, except that the source of the X-ray is installed in the chamber to have the beam incident on the sample, and a spherical mirror to detect the secondary electrons is positioned above the sample. The source of X-rays depends on the atoms to be detected. If GaAs(100) is used as a substrate, then Mg (1253.6 eV) or Al (1486.6 eV) K-line is used to detect the Ga line (1070 eV) and As line (1228 eV).

2.3.4.3. *What is measured*

The peak-to-peak intensity of the atoms is plotted against the thickness of the sample. A typical example of Co_2MnSi on GaAs(001) is shown in Fig. 2.10. By measuring the energy of the emitted secondary electron, the gray circle shown in Fig. 2.9, the chemical species can be determined.

2.3.4.4. *Remarks*

Because of the hole created in the core, AES can involve significant many-body interactions due to possible polarization effects induced by the hole. For simply determining the presence of certain species of atoms, however, it is possible to ignore such complications.

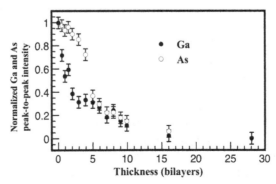

Fig. 2.10.   Peak-to-peak intensity as a function of thickness in Co_2MnSi at 450 K (Wang *et al.*, 2005b).

2.4. Methods of Determining Physical Properties

In this section, we focus on methods of determining magnetic, transport, and half-metallic properties. These methods are generally applied to all HMs.

2.4.1. *Magnetic properties*

With spintronic applications in mind, the important magnetic properties for a half-metallic sample are the saturation magnetization, magnetic anisotropy, spin polarization at E_F, polarization P, and the Curie temperature T_C. All of these properties have been studied both experimentally and theoretically. The experimental aspects will be discussed first.

Among all of these properties, the saturation magnetization is the most fundamental for FM properties. It is the saturation value of a hysteresis loop **B** as a function of **H**, where **B** is the magnetic induction and **H** is the applied magnetic field. For half-metallic compounds, it is important that this quantity is observed at or above RT in order to fabricate practical spintronic devices. Its value also can be compared to the theoretically predicted magnetic moment/unit cell. The methods used to measure a hysteresis loop include:

- Superconducting quantum interference device (SQUID)-based magnetometer.
- Technique utilizing the magneto-optical Kerr effect (MOKE).

2.4.1.1. *SQUID-based magnetometer*

The hysteresis loops of both bulk and thin-film samples can be measured by this method. Since SQUID — a sensitive device to measure the magnetic field — is commercially available, we shall not elaborate on the method.

Basically, a dc magnetic field **H** is applied along the easy axis of the sample. For thin-film samples, this axis, in general, lies in the plane of the film. The magnetization is measured as a function of the strength of the dc magnetic field.

Schematic setup In Fig. 2.11, a schematic setup of the measurement is shown. The large white area is the substrate, the sample is denoted by a small gray parallelogram. A dc magnetic field **H** is applied to the sample, indicated by an arrow. The SQUID is shown as a loop. It measures the **B** field from the sample.

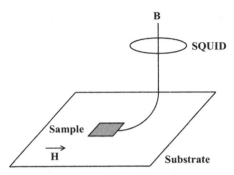

Fig. 2.11. A schematic setup for the SQUID magnetometer. A dc magnetic field **H** is applied to the sample. The SQUID is shown as a loop. It measures **B** field from the sample.

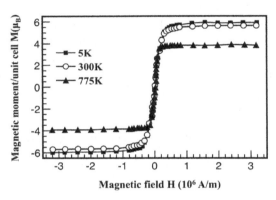

Fig. 2.12. Magnetic moment of Co_2FeSi as a function of **H** at different temperatures (Wurmehl *et al.*, 2005).

What is measured   In general, the hysteresis loop is measured. Often, the quantity **B** is converted into the magnetic moment. In Fig. 2.12, the magnetic moment of Co_2FeSi as a function of **H** is shown (Wurmehl *et al.*, 2005).

2.4.1.2. *Magneto-optical Kerr effect (MOKE)*

Basic idea   The Kerr effect occurs when the linear polarization of an electric field of incident light reflected from a magnetized sample is changed to elliptical polarization — the Kerr rotation. It was first applied to study surface magnetism by measuring the hysteresis loop of a magnetic thin film in the monolayer range (Bader, 1991). Since then, it has been used to search

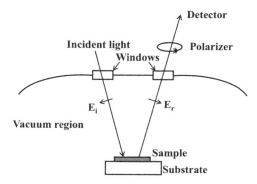

Fig. 2.13.   Schematic setup of MOKE.

for magnetic ordering, to identify major magnetic anisotropy, to correlate T_C to the layer thickness, and even characterize the critical exponent in a 2D phase transition.

Schematic setup   The setup is relatively simple: a schematic diagram is shown in Fig. 2.13. It can be incorporated into the growth vacuum chamber by adding an inlet for the incident light and an outlet for the reflected light if an external dc magnetic field **H** can be provided to the sample. If **H** is parallel to the film, it is called the longitudinal Kerr effect. The light source is a polarized light. The detection should include a polarizer.

What is measured   The main quantities to be measured are the intensity of the reflected light, and the Kerr rotation, which is proportional to the magnetic moment of the sample. By measuring the intensity of the reflected light, a hysteresis loop can be determined, and the coercivity can be extracted. The Kerr rotation itself does not provide the absolute value of the magnetic moment of the sample. The temperature dependence of the Kerr rotation is useful to determine the magnetization ratio $M(T)/M(0)$ — a relative quantity.

There are two ways to exhibit a hysteresis loop. One is to measure the Kerr rotation as a function of **H**. The shape of the loop and remanence are also functions of the layer thickness. The other way is to measure the intensity through the polarizer. With the polarization direction fixed in the polarizer, the intensity of the light passing through the polarizer changes as the strength of the dc magnetic field **H** varies. The two hysteresis loops are shown in Fig. 2.14.

(a) Kerr rotation   **(b) Kerr intensity**

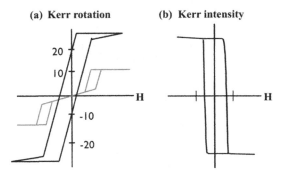

Fig. 2.14. (a) Hysteresis loop shown in Kerr rotation as a function of **H**. Results from different thickness are shown in black and light gray lines. Black loop is for thicker and light gray loop is for thinner samples. (b) Hysteresis loop shown in terms of the Kerr intensity as a function of **H** (Bader, 1991).

The sub-loops shown in light gray in Fig. 2.14(a) relate to the thickness of the sample. Let us take Co_2MnSi grown on GaAs(001) as an example. After growing three bilayers, the sample starts to exhibit an in-plane uni-axial magnetic anisotropy, with the easy axis pointing along the $[1\bar{1}0]$ direction. The experiment of measuring the hysteresis loop was carried out to probe the magnetization along the hard axis — the $[110]$ direction. Between 4 and 20 bilayers, there are two easy axes in the (110) plane. Hysteresis loops are developed, shown in gray in Fig. 2.14(a), as a function of positive and negative external magnetic fields. The remanence field is zero. As the film grows thicker — over 35 bilayers — the sub-loops disappear. The following empirical relation is found: let H_{s1} and H_{s2} be the fields in the middle of sub-loops at positive and negative **H**, then

$$H_s = (H_{s1} - H_{s2})/2. \tag{2.1}$$

The empirical relation is $H_s \approx 1/d$, where d is the thickness of the film. When H is applied along the easy axis, a simple hysteresis loop is obtained. In Fig. 2.14(b), the hysteresis loop is shown in terms of the Kerr intensity as a function of H.

The Kerr rotation under the remanence field along the easy axis can be measured as a function of temperature. If the Kerr rotation is assumed to be proportional to the magnetization, M, then the measured results fit well with the Bloch formula indicating the excitations are spin waves. The Bloch formula is

$$M(T) = M(0)(1 - bT^{3/2}). \tag{2.2}$$

Because MOKE cannot measure the absolute magnetization, it can only probe the magnetization in the surface region with penetration depth of the order of 10 nm. For Heusler alloys, this is a real disadvantage, because a complex layer stacking can happen to diminish magnetic properties at the surface. Alternative methods to measure the hysteresis loop are the alternative gradient magnetometer (AGM) and vibrating sample magnetometer (VSM). Since these two methods are specific to Heusler alloys, the related discussions will be given in Chapter 3.

2.4.1.3. *X-ray magnetic circular dichroism (XMCD)*

Samples like Co_2MnSi can have their moments contributed by multiple atomic species, such as Co and Mn. In order to determine individual contributions, XMCD can be used. The basic principle of the method has been discussed by Stöhr (1999).

Basic principles Within a one-electron model, an X-ray with either right- or left-hand circular polarization excites a core electron to a state above the vacuum level. Core states of a transition-metal element being excited normally have an orbital angular momentum of $l = 1$, the 2p core states. With the spin–orbit coupling, core states are identified as $L_3(l + s)$ and $L_2(l - s)$ lines, where s is the spin angular momentum. These lines are called L-edges. The XMCD signal is the normalized difference in absorption between the right- and left-hand polarized X-rays. It contains the information of exchange splitting and the spin–orbit coupling of the initial and final states. By controlling the energy of the X-ray, it is possible to have XMCD probe a particular element. After absorbing the X-ray, the core electron changes its angular momentum due to the circularly polarized light. It is excited to either an s- or a d-like state above the vacuum level. Spin-polarized bands respond differently for different polarizations of the X-ray. Measurements of photoelectrons with respect to the spin polarization of the sample provide information about the magnetic moment of each TM element in the sample. The processes are depicted in Fig. 2.15.

Schematic setup It is not easy to get a commercial X-ray source for frequencies in the desired range of a specific TM element. For Mn, energies of L_3 and L_2 lines are 638.5 eV and 651.4 eV, respectively. Tunable X-ray sources are available at Stanford Synchrotron Radiation Laboratory in the U.S., the Deutsches Elektronen-Synchrotron in Germany, and First Dragon

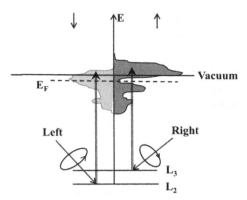

Fig. 2.15. Processes in XMCD. L_2 and L_3 are the core energy levels. The two arrows indicate the X-ray with the left and right polarization shown as circular arrows. vertical double arrows indicate possible transitions of core electrons. The spin-polarized density of states (DOS) are shown in shaded areas (Stöhr, 1999).

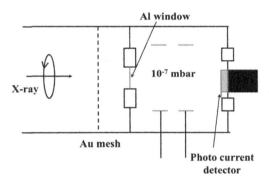

Fig. 2.16. Simple schematic setup of XMCD. X-ray is incident from the left. The dashed line is an Au mesh. The middle line at the right of the dashed line is the Al window. Horizontal lines are bias rings. The gray rectangle is the sample. The black box is the magnet with its south pole close to the sample and is separated from the sample by a quartz window. Emitted electrons are collected by the photocurrent detector (gray vertical line).

Beamline of NSRRC in Taiwan. With the light source, a schematic diagram of a typical setup is illustrated in Fig. 2.16.

Bias rings are used to eliminate photoelectrons from the gold mesh and stray electrons from the sample. The photocurrent detector measures the photoelectrons from the sample. The detector is made of materials which are transparent to the X-ray.

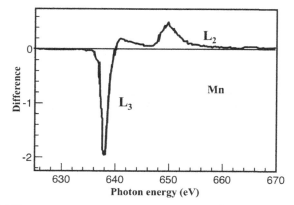

Fig. 2.17. XMCD at $L_{2,3}$ absorption edge of Mn for 45 Å thick Co_2MnSi films at RT (Wang *et al.*, 2005b).

What is measured The main quantities to be measured are the so-called white lines. They are the total photocurrents from L_2 and L_3 lines. By reversing the dc magnetic field, two sets of white lines are measured. The important magnetic circular dichroism (MCD) curves are obtained by subtracting two sets of white lines. A typical set of data for the difference of photocurrents associated with L_2 and L_3 lines (Wang *et al.*, 2005b) is shown in Fig. 2.17.

2.4.2. *Transport properties*

2.4.2.1. *Magnetic tunnel junctions (MTJs)*

Basic principles These junctions allow electrons to tunnel from one FM film through an insulating region to another FM film serving as a detector of spin polarization. The spin polarizations in such electrodes are defined by the DOS at E_F (Eq. (1.2)) and are denoted by P_1 and P_2, respectively. These junctions can be used to measure the resistance change if the magnetization of the detector is reversed by some controlled magnetic field.

Schematic setup The schematic setup is shown in Fig. 2.18. The two FM thin films are indicated by gray and black regions and are in contact with an ammeter denoted by an open polygon and a battery. The mid region between the gray and black regions is the insulating region.

What is measured Spin polarization of the sample is determined by measuring the tunnel magnetoresistance (TMR). The TMR can be

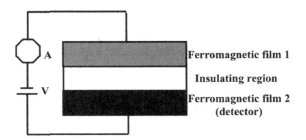

Fig. 2.18.   Schematic diagram of MTJs.

expressed in terms of spin polarizations:

$$\text{TMR} = \frac{2P_1 P_2}{1 - P_1 P_2}, \tag{2.3}$$

where P_1 and P_2 are spin polarizations at E_F in the electrodes. Note that it is also possible to measure directly the MR by reversing the magnetization in the detecting electrode. The MR is expressed as:

$$\text{MR} = \frac{R_{\uparrow\downarrow} - R_{\uparrow\uparrow}}{R_{\uparrow\uparrow}}, \tag{2.4}$$

where $R_{\uparrow\uparrow}$ and $R_{\uparrow\downarrow}$ are the resistance in parallel and antiparallel configurations in the two electrodes. In practice, measurements of TMR and MR are complicated by the presence of insulating layers. Cautions in interpreting the data are called for due to scatterings such as phonons.

These two kinds of results are given in Fig. 2.19. Figure 2.19(a) shows TMR as a function of bias voltage and Fig. 2.19(b) shows MR as a function of external field. The results also depend on the temperature and magnetization of electrodes. Figure 2.19(a) shows the RT case. The asymmetry shown in Fig. 2.19(a) originates from the tunneling process dependent on the direction of the bias voltage.

To see how the polarization can be determined, we follow the method suggested by Bezryadin *et al.* (1998) and Pesavento *et al.* (2004). Here we use the conductance instead of the resistance. For the parallel magnetization configuration of two electrodes, we expect that

$$G_{\uparrow\uparrow} = n_1 n_2 + (1 - n_1)(1 - n_2) \tag{2.5}$$

$$G_{\uparrow\downarrow} = n_1(1 - n_2) + (1 - n_1)n_2, \tag{2.6}$$

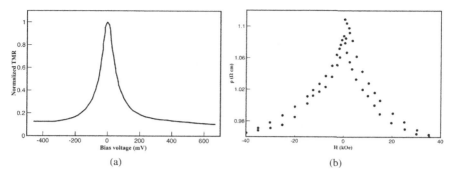

Fig. 2.19. (a) A typical plot of TMR as a function of bias voltage (Sakuraba *et al.*, 2006), and (b) a typical plot of MR as a function of external field (Gong *et al.*, 1997).

where $G_{\uparrow\uparrow}$ and $G_{\uparrow\downarrow}$ are conductances of the parallel and antiparallel configurations, respectively. n_1 is the fractional number of majority-spin electrons in the DOS at E_F in electrode 1, and n_2 is the corresponding fractional number of electrons in electrode 2. The number of minority-spin electrons is $1 - n_i, i = 1, 2$. $G_{\uparrow\uparrow}$ is the sum of the conductance from electrons with both electrodes having parallel orientations of majority- and minority-spin channels. $G_{\uparrow\downarrow}$ is for opposite spin orientations in the two electrodes. $G_{\uparrow\uparrow}$ and $G_{\uparrow\downarrow}$ are measured to determine n_1 and n_2. The polarizations in electrodes are:

$$P_1 = 2n_1 - 1 \tag{2.7}$$

$$P_2 = 2n_2 - 1. \tag{2.8}$$

2.4.2.2. *Resistivity*

General discussion For bulk materials, samples are often in rod form with the axis along the growth direction. For samples in thin-film form, the resistivity ρ is commonly measured in the plane of the sample. The resistivity is measured as a function of temperature. The temperature dependence is assumed to be:

$$\rho = \rho_o + cT^n, \tag{2.9}$$

where ρ_o is the residual resistivity — the low temperature limit. The exponent n is determined by fitting $\ln(\rho - \rho_o)$ as a function of $\ln(T)$. Three distinct temperature ranges were identified for the linear fits to Eq. (2.9) (Ritchie *et al.*, 2003). This information is used to find whether there is a

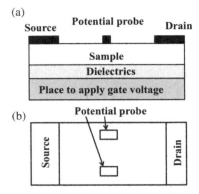

Fig. 2.20. A typical four-probe arrangement: (a) the side view and (b) the top view (Pesavento *et al.*, 2004).

structural or magnetic transition. At T_C, the resistivity shows a change in slope. In most cases, the slope changes from a mild slope, $n \approx 2$, to a steep one. The mild slope in the FM phase has been attributed to the absence of magnetic fluctuations; the sample behaves like an ordinary FM material. $n = 3$ indicates the contribution of phonons to the resistivity. A magnetic phase transition can be detected by the observed power-law shift.

Comments on setup   For both bulk and thin-film samples, the standard way of measuring resistivity is to attach four electrodes to the sample. This is called the "four-probe" arrangement. In Fig. 2.20, a typical four-probe arrangement is shown. One of the advantages of the four-probe arrangement is that it reduces the effects of the contact resistance between the sample and electrode (Bezryadin *et al.*, 1998; Pesavento *et al.*, 2004).

What is measured   The simplest measurement is ρ as a function of T. Typical data is shown in Fig. 2.21(a). In order to get the exponent n, the data are analyzed in terms of a plot of $\ln(\rho - \rho_o)$ as a function of $\ln(T)$. Figure 2.21(b) shows a typical plot of $\ln(\rho - \rho_o)$ as a function of $\ln(T)$.

2.4.2.3. *Hall conductivity*

Basic principles   The Hall effect is a result of the application of mutually perpendicular electric and magnetic fields on a conductor (Fig. 2.22). Moving charges experience a force, called the Lorentz force, when a magnetic field (H_z) is present that is perpendicular to their motion (I_x) and their paths are curved so that charges accumulate on the upper side of the

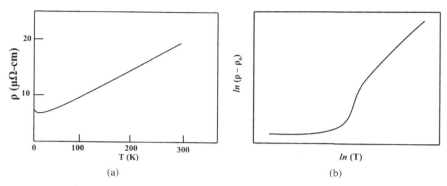

Fig. 2.21. (a) A typical plot of resistivity (ρ) as a function of temperature. (b) A typical plot of $\ln(\rho - \rho_o)$ as a function of $\ln(T)$.

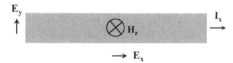

Fig. 2.22. The relative orientations of the magnetic field, current, and electric field. The sample is shown in gray. Current I_x flows in the x direction. The magnetic field H_z is in the z direction, into the paper (shown by the cross). The electric field E_y is in the y direction.

material. An electric field E_y is then developed along the y direction. When $E_y = -\omega_c \tau E_x$, where ω_c is the cyclotron frequency determined by H_z and τ is the lifetime of an electron when the moving electrons are no longer deflected, a voltage $V_H = E_y y$ is developed. This is the Hall effect, where V_H is called the "Hall voltage". The quantity of interest is called the "Hall coefficient" which is defined as:

$$R_H = V_H z / (I_x H_z), \qquad (2.10)$$

or

$$R_H = E_y / (j_x H_z), \qquad (2.11)$$

where $V_H = E_y y$, $I_x = j_x A_{yz}$, and A_{yz} is the cross-sectional area perpendicular to the current. The current density $\mathbf{J}$ in normal metals (NMs) or heavily doped semiconductors is given by

$$\mathbf{J} = \sigma_H \mathbf{H} \times \mathbf{E}, \qquad (2.12)$$

where σ_H is called the "Hall conductivity".

When Hall carried out his experiment on Fe, he found a coefficient about ten times larger than that of Ag. This effect in FM materials is now known as the "anomalous Hall effect" (AHE). Husmann and Singh (2006) expressed the current as:

$$\mathbf{J}' = \sigma'\mathbf{E} \times \mathbf{M}, \qquad (2.13)$$

where $\mathbf{J}'$ is the anomalous Hall current density, and σ' is the corresponding conductivity. The total current is the sum of the two currents, $\mathbf{J}$ and $\mathbf{J}'$. Physical origins of AHE have been classified into extrinsic and intrinsic mechanisms. The extrinsic mechanism includes spin–orbit interaction, impurities, and phonons. The intrinsic mechanism has been attributed to either the band structure (Yao *et al.*, 2004) or a topological spin background (Ye *et al.*, 1999).

Schematic setup   A schematic setup for a Hall measurement is shown in Fig. 2.23 (Moos *et al.*, 1995). The Hall voltage V_H is probed by the potential difference between electrodes at the two sides of the material. The two electrodes in the lower side of the material can improve the accuracy of V_H measurement. Initially, when $H = 0$, the resistor R is adjusted so that V_H is zero. The electrodes used to probe V_H need to have small cross sections in order to reduce stray current.

What is measured   Two quantities, the longitudinal σ_{xx} and the transverse σ_{xy} Hall conductivities as a function of H at different temperatures,

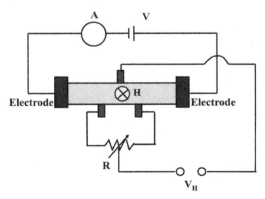

Fig. 2.23.   Schematic diagram for Hall measurement. The sample is shown as the long gray bar. The current is measured by the ammeter A. H is the strength of a magnetic field pointing into the paper (shown as a cross). V_H is the Hall voltage.

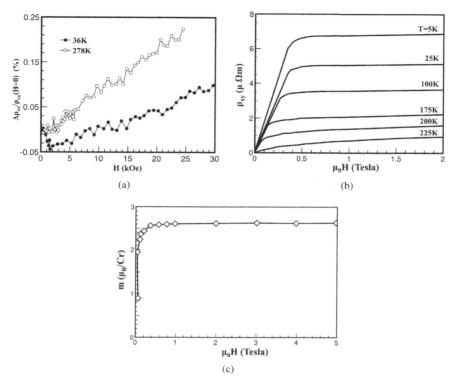

Fig. 2.24. (a) Longitudinal resistivity ratio ($\Delta\rho_{xx}/\rho_{xx}$) as a function of H (Husmann and Singh, 2006); (b) transverse resistivity (ρ_{xy}) as a function of H (Lee *et al.*, 2004); and (c) magnetization M, as a function of H (Lee *et al.*, 2004).

are commonly measured. Husmann and Singh (2006) measured $\Delta\rho_{xx}/\rho_{xx}$ as an alternative. The anomalous Hall coefficient R_{AH} is then extracted with the information from either longitudinal resistivity ratio $\Delta\rho_{xx}/\rho_{xx}$ as a function of H (Fig. 2.24(a)) or transverse resistivity ρ_{xy} as a function of H (Fig. 2.24(b)) measurements at different temperatures. For pure FM samples, R_{AH} is typically positive and monotonically decreasing with increasing temperature (Lee *et al.*, 2004). With disorder, R_{AH} can be negative.

2.4.3. *Half-metallic properties*

There are many predicted and grown half-metallic compounds which also exhibit FM properties. Therefore, one would expect that spin-polarized photoemission (SPP), ferromagnet-superconductor (SC) tunnel junction experiments, and the Andreev reflection method (Andreev, 1964) can be used to detect half-metallicity. The objective is to determine half-metallic properties of the samples and the spin polarization P at E_F. However, the first two experiments are not easily carried out. The energy resolution of the SPP method is needed within 10^{-3} eV, because only spin polarization at E_F is probed. The FM-SC tunnel junction requires a strictly uniform oxide layer formed on top of the SC such that the sample can be grown on top of the oxide layer for sensitive measurements of spin polarization. A more appealing approach to probe half-metallicity is the Andreev reflection method. This method has been applied to CrO_2 by Soulen *et al.* (1998) and Ji *et al.* (2001).

The SPP method is useful for determining the spin-polarized band structure. The FM-SC tunnel junction was used to measure spin polarization first. A more powerful approach, however, is the spin-polarized angle-resolved photoemission method (ARPES), which we discuss next.

2.4.3.1. *Spin-polarized angle-resolved photoemission spectroscopy (ARPES)*

Basic principles   A simple description of the geometry in an ARPES experiment is shown in Fig. 2.25(a). Light (the wavy line) with energy $\hbar\omega$ is incident on a solid. An electron (gray arrow) is then emitted. The energy of the emitted electron satisfies the following conditions depending

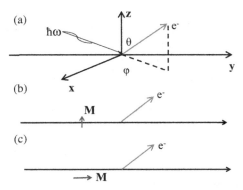

Fig. 2.25.   Geometric configurations for (a) incident light and angle-resolved photoemission, (b) longitudinal magnetization, and (c) transverse magnetization. θ and ϕ in (a) are the polar and azimuthal angles.

on whether the electron originates at E_F or from a valence state below E_F.

$$E_{\text{kin}} = \hbar\omega - \Phi \qquad (2.14)$$

$$E_{\text{kin}} = \hbar\omega - \Phi - E_b, \qquad (2.15)$$

where E_{kin} is the maximum energy of the emitted electron from a state at E_F, $\hbar\omega$ is the energy of the incident light, Φ is the work function, and E_b is the binding energy with respect to E_F. The momentum relations, between the electron inside and electron outside the sample, determining the angle-resolved spectra are

$$\mathbf{k}_{\parallel}^{\text{out}} = \mathbf{k}_{\parallel}^{\text{in}} \pm \mathbf{G} \qquad (2.16)$$

$$k_{\perp}^{\text{out}} = \sqrt{\frac{2mE_{\text{kin}}}{\hbar^2} - |\mathbf{k}_{\parallel}^{\text{out}}|^2}, \qquad (2.17)$$

where $\mathbf{k}_{\parallel}^{\text{out}}$ is the momentum of the emitted electron parallel to the surface of the sample, $\mathbf{k}_{\parallel}^{\text{in}}$ is the parallel component of the momentum of the electron inside the sample, $\mathbf{G}$ is a reciprocal lattice vector parallel to the surface, and $k_{\perp}^{\text{out}}$ is the magnitude of the momentum of the emitted electron perpendicular to the surface. Both $\mathbf{k}_{\parallel}^{\text{out}}$ and $k_{\perp}^{\text{out}}$ can be specified by the polar and azimuthal angles (θ, ϕ) shown in Fig. 2.25(a).

An electron excited by the incident light can be involved in many-body interactions as it travels toward the surface. The effects of these interactions can be discerned by comparing measured spectra to band structure results (Braun, 1996).

Remarks To probe electronic states in an FM sample using photoemission process, one must distinguish the two distinct polarizations associated with spins of electrons. It was not until 1979, with the work of Celotta *et al.*, that significant advances were made experimentally. Earlier, only the geometry of the longitudinal magnetization (Fig. 2.25(b)) was considered. In this geometry, a large external magnetic field is required to align domains of the sample in such a way that the magnetization is perpendicular to the surface. Consequently, the resultant magnetic field can deflect the emitted electrons — making it impossible to carry out angle-dependent photoemission measurements.

Celotta *et al.* suggested the use of a long, thin-plate form of sample, with the magnetization parallel to the surface of the sample without applying an external field. In the transverse magnetization experiment

(Fig. 2.25(c)), the sample is permanently magnetized. The stray field outside the sample is in general too small to have a significant effect on the trajectory of the emitted electron. One may also employ a particular electron extraction lens system to ensure that emitted electrons will not be affected by the presence of the stray field.

The emitted electrons can have two spin polarizations with respect to a chosen axis. To detect net polarizations, one carries out double scattering measurements. The first scattering selects a particular spin polarization. The next step utilizes the scattering process proposed by Mott to determine the cross section of each spin polarization. This is called "Mott scattering".

Mott scattering   In its original form, Mott scattering characterizes the Coulomb scattering cross section modified by exchange effects when two equivalent electrons scatter (Bohm, 1951). To understand spin-polarized photoemission, Mott scattering is applied to an electron having its spin momentum coupled to its orbital angular momentum when the electron is scattered by the Coulomb interaction from a target composed of a heavy element, such as gold (Au). The effect of this spin–orbit interaction causes a beam of electrons with a particular spin orientation passing at the left and right of the target to show asymmetry. We expect that a new generation of light source for photoemission experiments will be available in the near future; it is therefore worth providing the background of the spin–orbit interaction. We start with the Dirac equation in 2×2 matrix form (Kessler, 1985).

$$[E - e\Phi - c\alpha \cdot (\mathbf{p} - e\mathbf{A}/c) - \beta mc^2]\psi = 0, \tag{2.18}$$

where E is the energy of the electron, Φ is the external potential, and α and β are 2×2 matrices

$$\alpha_i = \begin{bmatrix} 0 & \sigma_i \\ -\sigma_i & 0 \end{bmatrix}; \quad \beta = \begin{bmatrix} 1 & 0 \\ 0 & 1 \end{bmatrix} \tag{2.19}$$

where i $= 1, 2, 3$ or x, y, z, and

$$\sigma_1 = \begin{bmatrix} 0 & 1 \\ 1 & 0 \end{bmatrix}, \quad \sigma_2 = \begin{bmatrix} 0 & -i \\ i & 0 \end{bmatrix}, \quad \sigma_3 = \begin{bmatrix} 1 & 0 \\ 0 & -1 \end{bmatrix}, \tag{2.20}$$

$\mathbf{p}$ is the momentum of the electron, $\mathbf{A}$ is the vector potential for externally applied magnetic and electric fields, Φ is the Coulomb potential from the nuclei and m is the mass of the electron. By expressing Eq. (2.18) in a 2×2 matrix equation and the wave function into large and small components with respect to the rest mass, the resultant equation of the large component wave function is the 2×2 Schrödinger equation:

$$\left\{ \frac{1}{2m} \left(\mathbf{p} - \frac{e\mathbf{A}}{c} \right)^2 + e\Phi - \frac{e\hbar}{2mc}\sigma \cdot \mathbf{B} + i\frac{e\hbar}{4m^2c^2}\mathbf{E} \cdot \mathbf{p} - \frac{e\hbar}{4m^2c^2}\sigma \cdot \mathbf{E}_C \times \mathbf{p} \right\} \psi$$

$$= E_s\psi, \tag{2.21}$$

where the first two terms are the usual Schrödinger equation for an electron in the presence of an external field. The third term is the Zeeman term from the spin moment $\mu = \frac{e\hbar}{2mc}\sigma$ of the electron in the external magnetic field $\mathbf{B}$. The fourth term is from an external electric field. The last term can be reduced to the spin–orbit interaction. In the electron reference frame, the electron with velocity $\mathbf{v}$ sees the nucleus of the heavy atom moving at $-\mathbf{v}$. The magnetic field $\mathbf{B}_C$, due to motion of the nucleus and the Coulomb interaction from the nucleus (the current associated with the nucleus), is

$$\mathbf{B}_C = -\mathbf{v} \times \frac{\mathbf{E}_C}{c} = \frac{\mathbf{E}_C \times \mathbf{p}}{mc}, \tag{2.22}$$

and $\mathbf{E}_C$ is the electric field from the Coulomb potential, V.

$$\mathbf{E}_C = -\frac{1}{e}\frac{dV}{dr}\frac{\mathbf{r}}{r}. \tag{2.23}$$

Now, the magnetic moment of the electron interacts with $\mathbf{B}_C$. With $\mathbf{s} = \frac{\hbar}{2}\sigma$, Eqs. (2.22), and (2.23), the last term in Eq. (2.21) can be expressed as:

$$-\mu \cdot \mathbf{B}_C = \frac{e}{2m^2c^2}\mathbf{s} \cdot \left[-\frac{1}{e}\frac{dV}{dr}\frac{\mathbf{r}}{r} \times \mathbf{p} \right] = \frac{1}{2m^2c^2}\frac{1}{r}\frac{dV}{dr}(\mathbf{l} \cdot \mathbf{s}), \tag{2.24}$$

where $\mathbf{l} = \mathbf{r} \times \mathbf{p}$.

The effects of this term can be seen as follows: let an electron pass to the right of the target (an Au atom). The orbital angular moment is oriented along the z-axis in the frame of the moving electron. When the electron occupies the ↑ spin state, with orientation aligned along the z-axis, it experiences an additional positive potential from Eq. (2.24) with respect to the Coulomb potential. This resultant potential $V_\uparrow$ is shown as a solid curve in Fig. 2.26. The Coulomb potential is shown as a dashed curve. Similarly, the ↓ spin electron experiences an additional negative potential and the resultant potential $V_\downarrow$ is shown as a dash-dot curve in Fig. 2.26.

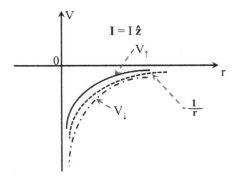

Fig. 2.26. Different potentials experienced by the ↑ and ↓ spin states in Mott scattering.

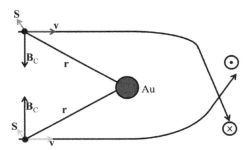

Fig. 2.27. An electron passing to the left (top) (dark gray arrows for the spin and velocity) and right (bottom) (light gray arrows) of an Au atom experiences different spin–orbit interactions due to the reversal of the effective $\mathbf{B}_C$ field. $\mathbf{v}$ is the velocity of the electron and $\mathbf{S}$ is the spin of the electron.

Different potentials for the ↑ and ↓ spin states result in different cross sections. If an electron passes to the right side of the target, the orbital angular momentum reverses direction, so that the extra potential due to the last term acting on the ↑ and ↓ spin states is also reversed. The phenomenon is depicted schematically in Fig. 2.27.

Consequently, when a beam of electrons with a definite spin polarization scatter from a target of heavy atoms, there is an asymmetry recorded in the detectors located at the left and right of the target. This is the basic idea behind the Mott detector.

As discussed above, the setups can involve delicate lens systems. One of the more detailed setups is designed by Raue *et al.* (1984) and enables one to determine the energy and angular resolution of spin-polarized photoelectrons. A schematic diagram of the setup is shown in Fig. 2.28(a),

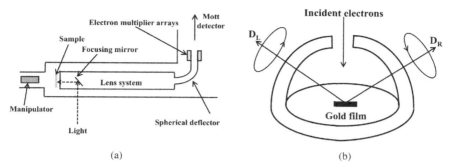

(a)   (b)

Fig. 2.28. (a) Schematic of a photoemission experiment. (b) A retarding-potential-type Mott detector.

in which a vacuum chamber contains an FM sample, manipulator, mirror with focusing system, and other electron lens systems. At the end of the chamber, a 90° spherical deflector deflects photoelectrons into the Mott detector or Mott spin polarimeter by passing through another set of lens systems. The spherical deflector serves also as an energy analyzer. At its exit plane, an assembly of electron multiplier arrays are installed for multichannel detections.

The Mott detector is based on the principle of Mott scattering (see page 42). The detector is either a conventional high-energy type used by Getzlaff *et al.* (1998) or a retarding-field type used by Uhrig *et al.* (1989). The former exhibits a drawback of poor discrimination against inelastic scattering incurred by electrons. The retarding-field-type Mott detector consists of an energy analyzer and electron deflectors. It has the capability to discriminate the energy of electrons and monitor the polarization of an electron even with spatial asymmetry caused by the spin–orbit interaction. However, a slight asymmetric alignment of the instrument can smear the polarization measurements. A way to eliminate this smear is to include measurements with reversed spin polarizations of electrons. A retarding-field-type Mott detector is shown in Fig. 2.28(b). The outer two hemispheres provide the acceleration to incoming photoelectrons. The electrons scatter from the gold film with the spin–orbit interaction. The scattered electrons enter the left detector (D_L) and the right detector (D_R). Both detectors are equipped with electron multipliers to increase the sensitivity of measurements.

The operational details are given below: the asymmetry is defined as the ratio of the left current I_L and right current I_R with respect to the

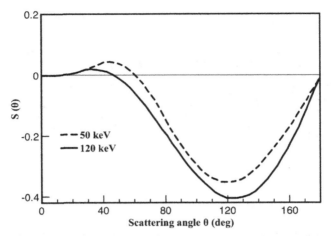

Fig. 2.29. Sherman function as a function of θ, the dashed curve is for the electronic energy at 50 keV and the solid curve is for 120 keV (Gellrich and Kessler, 1991).

target

$$A = \frac{I_L}{I_R}. \tag{2.25}$$

The polarization in the Mott detector is given as

$$P_{\text{Mott}} = S_{\text{eff}} P = \frac{A - 1}{A + 1}, \tag{2.26}$$

where S_{eff} is called the Sherman function (Fig. 2.29) and

$$P = \frac{N_\uparrow - N_\downarrow}{N_\uparrow + N_\downarrow}. \tag{2.27}$$

The Sherman function is related to the scattering amplitude and phase shifts, which can be seen as follows: let us first assume that the wave function of the $\uparrow$ spin channel after the scattering is given asymptotically $(r \to \infty)$ by

$$\psi^\uparrow \to e^{ikz} + \frac{f_1(\theta, \phi)e^{ikr}}{r}. \tag{2.28}$$

Because of the spin–orbit interaction, there is a possibility of having a spin-flip transition,

$$\psi^{\uparrow \to \downarrow} \to \frac{g_1(\theta, \phi)e^{ikr}}{r}. \tag{2.29}$$

Similarly for the $\downarrow$ spin channel with quantization axis along the **z**-direction, we have

$$\psi^{\downarrow} \rightarrow e^{ikz} + \frac{f_2(\theta, \phi)e^{ikr}}{r}. \tag{2.30}$$

with the spin-flip transition,

$$\psi^{\downarrow \rightarrow \uparrow} \rightarrow \frac{g_2(\theta, \phi)e^{ikr}}{r}. \tag{2.31}$$

The amplitudes f_i and g_i, $i = 1, 2$, can be expanded into partial waves with the corresponding phase shifts η_l and η_{-l-1}. The scattered waves f_i and g_i can be written as

$$f_1(\theta, \phi) = \frac{1}{2ik} \sum_{l=0}^{\infty} \{(l+1)(e^{2i\eta_l} - 1) + l(e^{2i\eta_{-l-1}} - 1)\} P_l(\cos\theta)$$

$$= f(\theta) = f_2(\theta, \phi), \tag{2.32}$$

$$g_1(\theta, \phi) = \frac{1}{2ik} \sum_{l=0}^{\infty} \{-e^{2i\eta_l} + e^{2i\eta_{-l-1}}\} P_l^1(\cos\theta)e^{i\phi}$$

$$= g(\theta)e^{i\phi}, \tag{2.33}$$

and

$$g_2(\theta, \phi) = -g(\theta)e^{-2i\phi}, \tag{2.34}$$

where $P_l(cos\theta)$ is the Legendre polynomial and $P_l^1(cos\theta)$ is the associated Legendre polynomial. For an incident wave with an arbitrary spin orientation, the spinor is:

$$\psi = A \begin{bmatrix} 1 \\ 0 \end{bmatrix} e^{ikz} + B \begin{bmatrix} 0 \\ 1 \end{bmatrix} e^{ikz}. \tag{2.35}$$

$$\begin{bmatrix} a \\ b \end{bmatrix} \frac{e^{ikr}}{r} = A \begin{bmatrix} f_1 \\ g_1 \end{bmatrix} \frac{e^{ikr}}{r} + B \begin{bmatrix} g_2 \\ f_2 \end{bmatrix} \frac{e^{ikr}}{r} = \begin{bmatrix} Af - Bge^{-2i\phi} \\ Bf + Age^{i\phi} \end{bmatrix} \frac{e^{ikr}}{r}. \tag{2.36}$$

The cross section is

$$\sigma(\theta, \phi) = \frac{|a|^2 + |b|^2}{|A|^2 + |B|^2} = (|f|^2 + |g|^2) + (fg^* - f^*g)\frac{-AB^*e^{i\phi} + A^*Be^{-i\phi}}{|A|^2 + |B|^2}$$

$$= (|f|^2 + |g|^2)(1 + S(\theta))\frac{-AB^*e^{i\phi} + A^*Be^{-i\phi}}{|A|^2 + |B|^2}. \tag{2.37}$$

$S(\theta)$ is defined as

$$S(\theta) = \frac{(fg^* - f^*g)}{|f|^2 + |g|^2}. \tag{2.38}$$

It does not depend on ϕ and it can be determined by spin-unpolarized photoemission experiments.

Since the Mott detector is sensitive to misalignment of the incident beam and electrons backscattered from walls, the experiments and interpretations of the results must be carried out with great care.

What is measured   The angle-resolved energy distribution curve (EDC) ($I_0 = I_\uparrow + I_\downarrow$) and spin-polarized EDC ($P = (I_\uparrow - I_\downarrow)/(I_\uparrow + I_\downarrow)$) are measured simultaneously (Raue *et al.*, 1984). From these two measurements, the two spin-resolved contributions $I_\uparrow$ and $I_\downarrow$ can be calculated as shown in the lower part of Fig. 2.30. Experimental results on HMs, CrO_2 in particular, will be discussed in Chapter 4.

In the following, we present two schemes for probing spin-polarized electrons.

2.4.3.2. *Ferromagnet-superconductor tunneling*

Basic principles   Meservey and Tedrow (1994) were the first to carry out the experiment on a Al–Al_2O_3–Ni junction, where Al is in a thin-film form and exhibiting superconducting properties. It demonstrates the spin-polarized electron tunneling effects. The density of states of Al and Ni are schematically drawn in Fig. 2.31(a). Without the external magnetic field, the density of states of Al are denoted by solid curves. Because of the

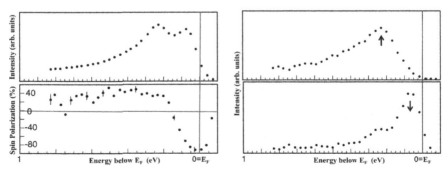

Fig. 2.30.   Left panels: EDC and spin polarization for a normal emission from Ni(110). Right panels: calculated spin-resolved EDCs (Raue *et al.*, 1984).

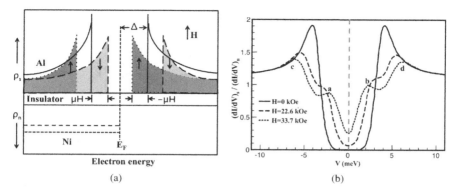

Fig. 2.31. (a) The Al (superconducting) density of states denoted by solid curves are shown in the upper panel. Δ is half of the gap. The middle horizontal region is an insulating barrier. The bottom panel illustrates the Ni density of states. When an external magnetic field H is applied paralleled to the Al thin film, the SC exhibits Zeeman splitting. Dashed curves are for the $\downarrow$ spin states while dotted curves are for the $\uparrow$ spin states. (b) Ratio of conductance as a function of voltage for different values of H (Meservey and Tedrow, 1994).

superconducting gap, 2Δ is small ($\approx k_B T$), Ni density of states are approximated by constants, with the $\uparrow$ spin states marked by the short dashed line and $\downarrow$ spin states marked by the dashed line. The middle region is an insulating barrier. When an external magnetic field **H** is applied parallel to the film, the Al density of states are Zeeman split (shown as the dashed ($\downarrow$ spin state) and dotted ($\uparrow$ spin state) curves).

For a finite H and an NM used as an electrode, the tunneling current is expected to be symmetric with respect to spin polarization as the voltage reverses. If the non-superconducting electrode is a ferromagnet, then an asymmetry is expected for the $\uparrow$ spin and $\downarrow$ spin peaks. An expected tunneling conductance is shown in Fig. 2.31(b). The asymmetry is indicated by a, b, c, d peaks at $H = 22.6$ and $33.7\,$kOe.

What is measured The tunneling conductance, dI/dV, as a function of the bias voltage, V, is measured. Figure 2.31(b) shows the typical results. In practice, effects of the interface and spin–orbit interaction for heavy elements should be dealt with.

2.4.3.3. *Andreev reflection*

Basic principles The basic principle of the Andreev reflection method is illustrated in Fig. 2.32. It consists of the density of states involved in the

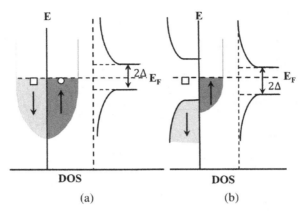

Fig. 2.32. (a) Energy as a function of DOS of tunneling between an NM and an SC. (b) Energy as a function of DOS of tunneling between an HM and an SC. Arrows mark the spin polarizations of states in the metal.

tunneling experiments of NM/SC (Fig. 2.32(a)) and HM/SC (Fig. 2.32(b)). It is one of the methods providing unambiguous determination of a sample to be an HM.

For the NM/SC junction (Fig. 2.32(a)), both majority- (↑) and minority- (↓) spin states are occupied at E_F. If a majority-spin electron is incident (under a bias voltage, V, less than Δ) to the SC region, a majority-spin hole (open circle) will contribute to the reflected current. At the same time, a minority-spin electron with proper momentum forms a Cooper pair with the incident electron and leaves behind a minority-spin hole in the metal region (shown as an open square in Fig. 2.32(a)). The extra current caused by this hole is the Andreev reflection. The conductance is contributed by two electrons passing through the junction. If the NM is replaced by an FM metal, a similar process applies even though there are more majority-spin electrons than minority-spin electrons. On the other hand, if an HM is considered, the tunneling carriers from the HM side have only one spin orientation. It is impossible for the incident carrier entering the SC to pair with the minority-spin electron due to the presence of the gap (shown as an open square in the HM region). As a result, a reflection by the hole generated by the electron in the insulating channel can not appear in the HM. The conductance is therefore zero. To apply this method to FM HMs, the junction is in a point-contact form so the uniform oxide layer is not needed.

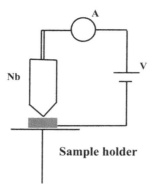

Fig. 2.33. Schematic of Andreev reflection method for detecting half-metallic properties. The gray rectangle is the sample. The Nb metal is in the tapered form serving as a point contact. The circle "A" is the ammeter. "V" represents the battery.

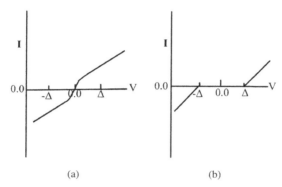

Fig. 2.34. (a) The tunneling I-V curve between an NM and an SC. (b) The tunneling I-V curve between an HM and an SC.

Schematic diagram of setup Niobium (Nb) metal is commonly used as the SC in the Andreev reflection method. It is in the form of a tip. A schematic diagram of the setup is shown in Fig. 2.33.

What is measured The tunneling current, from the reading of the ammeter, is the crucial quantity to be measured. To contrast characteristics relevant to processes of the NM-SC tunneling and HM-SC tunneling, the corresponding tunneling currents as a function of voltage are shown in Fig. 2.34. Note that in Fig. 2.34(a), the current is zero only when the bias voltage is zero. However, there is a range of 2Δ of the voltage for the tunneling current to be zero in Fig. 2.34(b).

A phenomenological approach to the tunneling of spin-polarized electrons has been given by Blonder, Tinkham and Klapwijk (1982).

Analysis   Blonder *et al.* introduced a model repulsive potential $V_{int}\delta(x)$ at the interface to explain the NM-SC tunneling. One of the important parameters used in the model is the normalized barrier strength Z $(= V_{int}/(\hbar v_F))$, where v_F is the Fermi velocity, to characterize the interface and effects of a barrier. Blonder *et al.* considered the current passing through a plane in the NM region since it is contributed by single particles instead of Cooper pairs. Furthermore, contributions to the current are parametrized by four probabilities A, B, C, and D depending on (E, Δ, Z) and the tunneling between NM and nonmagnetic or between NM and SC, where E is the energy of the incident electron and 2Δ is the gap of the SC. The simplified notations are $A(E), B(E), C(E)$, and $D(E)$. They characterize probabilities for the four processes.

To see the four processes described by Blonder *et al.*, a simplified band diagram is shown in Fig. 2.35 for an NM or FM metal and an SC in contact. At equilibrium, quasiparticle states labeled 0 and 5 at left and 1 to 4 at right are occupied with the same probability $f_0(E)$. There is a hole state at left labeled by 6. The lines for momenta in the two regions are taken as positive when they are on the right-hand side of energy axes and negative when they are on the left-hand side. Consider now an electron labeled by "0" with momentum k_{in} and $E > \Delta$ incident from the NM side into the SC. By matching the slope and value of the wave function at the interface, the

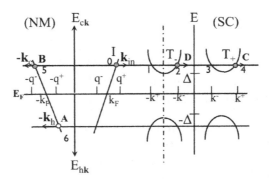

Fig. 2.35.   The band diagram of an NM or FM metal and an SC in point contact (Blonder *et al.*, 1982). 2Δ is the gap of the SC. I (small filled circle) is the incident electron. T_- and T_+ are the possible transmitted electronic states. The hole state labeled by A is for the Andreev reflection.

incident electron can be subjected to transmission and reflection processes labeled by A, B, C, and D. For example, $C(E)$ is the normal transmission probability through the interface without change of sign of momentum $q^+ = k^+$. $D(E)$ is the transmission probability for $q^+ = -k^-$ crossing the Fermi surface. $A(E)$ characterizes the Andreev reflection probability for a hole on the other side of the Fermi surface. A hole appearing in the NM region is required for the incident electron to form a Cooper pair in the SC. $B(E)$ is the probability for an ordinary reflection. The four parameters are obtained by solving Bogoliubov equations. These four parameters satisfy the sum rule for the conservation of one electron:

$$A(E) + B(E) + C(E) + D(E) = 1. \tag{2.39}$$

The tunneling current I is first expressed in terms of the difference of the right-going $f_\rightarrow(E)$ and left-going $f_\leftarrow(E)$ distribution functions,

$$I = 2eA_\alpha N_\sigma(E_F)v_{F\sigma} \int_{-\infty}^{\infty} [f_\rightarrow(E) - f_\leftarrow(E)]dE, \tag{2.40}$$

where A_α is the effective cross section of the contact, $N_\sigma(E_F)$ is the DOS at E_F for the σ spin state, $v_{F\sigma}$ is the Fermi velocity for the σ spin state, and E is the energy of the incident electron. For complicated cases, the two distribution functions must be solved self-consistently (Blonder *et al.*, 1982). Consider a situation:

$$f_\rightarrow(E) = f_0(E - eV) \tag{2.41}$$

and

$$f_\leftarrow(E) = A(E)(1 - f_\rightarrow(-E)) + B(E)f_\rightarrow(E) + (C(E) + D(E))f_0(E), \tag{2.42}$$

with

$$f_\rightarrow(-E) = f_0(E + eV), \tag{2.43}$$

where f_0 is the Fermi–Dirac distribution function and V is the potential energy due to an externally applied voltage. Substituting Eqs. (2.41) and (2.42) into Eq. (2.40), the current from the NM to the SC is,

$$I_{NM \rightarrow SC} = 2eA_\alpha N_\sigma(E_F)v_{F\sigma} \int_{-\infty}^{\infty} \{f_0(E - eV) - f_0(E)\}$$
$$\times \{1 + A(E) - B(E)\}dE. \tag{2.44}$$

From this expression, the conductance can be obtained by taking the derivative with respect to V. For an NM-NM contact, $A(E)$ is zero because there is no Andreev reflection and $D(E)$ is also zero due to no cross-Fermi-surface scattering. With Eq. (2.42) and the sum rule Eq. (2.39), the conductance is

$$\frac{dI_{\text{NM}-\text{NM}}}{dV} = 2eA_\alpha N_\sigma(E_F)v_{F\sigma}C(E) = G_n. \tag{2.45}$$

G_n is also the conductance for the NM-SC contact with $V \gg \Delta$. The integrations in Eq. (2.44) were carried out for $Z = 0.0$, 0.5, 1.5, and 5.0, respectively. The results are shown in Fig. 2.36. For the $Z = 0$ case, the normalized conductance is 2 for $V < \Delta$. This is due to the presence of a Cooper pair and G_n should correspond to one electron. For $|V| \gg \Delta$, we have

$$\frac{G(V)}{G_n} = 1. \tag{2.46}$$

To see how to extract the polarization information, we consider the definition of the spin polarization. It is known for the tunneling junction that the spin polarization P_T can be determined by

$$P_T = \frac{N_\uparrow(E_F)|T_\uparrow|^2 - N_\downarrow(E_F)|T_\downarrow|^2}{N_\uparrow(E_F)|T_\uparrow|^2 + N_\downarrow(E_F)|T_\downarrow|^2}, \tag{2.47}$$

where T_σ is the tunneling matrix element for the σ spin channel.

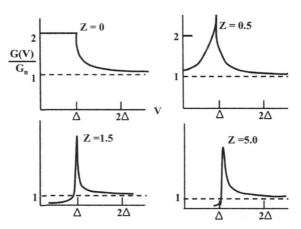

Fig. 2.36.   Numerical $G(V) = dI/dV$ normalized to G_n as a function of V for $Z = 0.0$, 0.5, 1.5, and 5.0, respectively (Blonder *et al.*, 1982).

For the point contact, it is not the tunneling matrix elements which govern the process but the speed of electrons. The spin polarization for a point contact P_c is expressed as

$$P_c = \frac{N_\uparrow(E_F)v_{F\uparrow} - N_\downarrow(E_F)v_{F\uparrow}}{N_\uparrow(E_F)v_{F\uparrow} + N_\downarrow(E_F)v_{F\uparrow}}. \tag{2.48}$$

Because the product $N_\sigma(E_F)v_{F\sigma}$ is the current I_σ for σ channel, P_c can be rewritten in terms of currents:

$$P_c = \frac{I_\uparrow - I_\downarrow}{I_\uparrow + I_\downarrow}. \tag{2.49}$$

This expression is based on the assumption of ballistic transport (i.e., no scattering) at the point contact. In general, P_c should relate to the parameter Z introduced by Blonder *et al.* to characterize the scattering at the contact. Strijkers *et al.* (2001) modified the current expression into the sum of fully polarized current (I_{pol}) and fully unpolarized current (I_{unpol}) for the NM and SC contact,

$$I = (1 - P_c)I_{\text{unpol}} + P_cI_{\text{pol}}, \tag{2.50}$$

where $(1 - P_c)I_{\text{unpol}}$ is the unpolarized part of the current and can include a contribution from the Andreev reflection and P_cI_{pol} is the polarized part of the current and does not have any contribution from the Andreev reflection. The conductance can now be related to P_c by:

$$G(V) = \frac{dI(V,T;P_c,Z)}{dV} = (1 - P_c)\frac{dI_{\text{unpol}}(V,T;Z)}{dV} + P_c\frac{dI_{\text{pol}}(V,T;Z)}{dV}. \tag{2.51}$$

P_c is obtained by fitting Eq. (2.51) to the normalized conductance from experiments with different Z values (Fig. 2.36). Figure 2.36 shows that the conductance relevant to the Andreev reflection decreases as Z increases.

2.4.3.4. *Curie temperature T_C*

General comments In order to ensure that a device made of an HM can be operated at RT, the determination of T_C of an HM is essential. There are several methods to determine T_C of an HM. A commonly used one is to measure hysteresis loops at different temperatures. An alternative method is to measure the saturation magnetization as a function of temperature. Measurements of hysteresis loops have been given in Section 2.4.1. Here, we shall briefly describe measurements of the saturation magnetization.

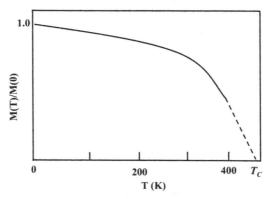

Fig. 2.37.   Saturation magnetization as a function of temperature. T_C is obtained by extrapolation.

Saturation magnetization   SQUID is commonly used to measure the saturation magnetization. It is a very sensitive way to determine the magnetization. The data are normalized to the magnetization at zero temperature. A typical set of data is shown in Fig. 2.37. T_C is then obtained by extrapolation.

2.5. Theoretical Methods

The electronic DOS and magnetic moment of half-metallic compounds are the primary interest. In addition, energetics for stable structures and the effects of dopings on half-metallic properties at $T = 0\,\mathrm{K}$ are of interest. Theoretical methods based on DFT (Hohenberg and Kohn, 1964) in the local density approximation (LDA) (Kohn and Sham, 1965) or generalized gradient approximation (GGA) (Langreth and Perdew, 1980; Langreth and Mehl, 1983) for exchange-correlation are the most regularly used. We shall give a brief review of DFT and comment on popular methods of calculating the total energy of HMs and corresponding electronic band structures.

2.5.1. *Density functional theory (DFT)*

The basic idea of DFT is that in a many-electron system the ground state is assumed to be nondegenerate and the charge density $\rho(\mathbf{r})$ is the fundamental quantity. If ρ is known, other physical properties of the many-electron system can be determined. There are many reviews of DFT in the physics and chemistry literature. We shall not dwell on details of the theory but will just provide an overview of the physical meaning and important consequences. The theory is based on two theorems by Hohenberg and Kohn (1964).

2.5.1.1. *Hohenberg–Kohn theorem I*

There is a *one-to-one* correspondence between the ground state charge density $\rho(\mathbf{r})$ of a many-electron system and external potential V_{ext}. As a consequence of this theorem, there is a unique ground state charge density for any given external potential acting on a many-electron system.

2.5.1.2. *Hohenberg–Kohn theorem II*

The ground state charge density of a many-electron system minimizes the total energy of the system. Mathematically, the total energy functional F of a many-electron system at its ground state is composed of the kinetic energy T, external potential energy $V_{\text{ext}}[\rho]$, and interaction energy of electrons. This interaction energy can be divided into the Coulomb interaction between any pair of electrons and exchange-correlation energy E_{xc}. These energies are functionals of the charge density, which is a function of the electronic position $\mathbf{r}$. Let the total energy functional of the system be F,

$$F = \int T[\rho]d^3r + V_{\text{ext}}[\rho] + \frac{1}{2}\iint \frac{\rho(\mathbf{r})\rho(\mathbf{r}')}{|\mathbf{r}-\mathbf{r}'|}d^3r d^3r' + \int E_{\text{xc}}[\rho]d^3r, \tag{2.52}$$

where $V_{\text{ext}}[\rho]$ is expressed in terms of a single particle potential $v_{\text{ext}}(\mathbf{r})$,

$$V_{\text{ext}}[\rho] = \int \rho(\mathbf{r})v_{\text{ext}}(\mathbf{r})d^3r. \tag{2.53}$$

The free energy functional F of the system is minimized with respect to $\rho(\mathbf{r})$ with the constraint,

$$N = \int \rho(\mathbf{r})d^3r. \tag{2.54}$$

N is the total number of electrons in the system. By introducing a Lagrange multiplier λ in Eq. (2.52) to enforce Eq. (2.54) a one-particle equation is obtained through the minimization of F with respect to $\rho(\mathbf{r})$,

$$t[\rho(\mathbf{r})] + v_{\text{ext}}(\mathbf{r}) + \int \frac{\rho(\mathbf{r}')}{|\mathbf{r}-\mathbf{r}'|}d^3r' + V_{\text{xc}}[\rho(\mathbf{r})] - \lambda = 0, \tag{2.55}$$

where $t[\rho(\mathbf{r})] = \delta T/\delta\rho$ and $v_{xc}[\rho(\mathbf{r})]$ is $\delta E_{xc}[\rho(\mathbf{r})]/\delta\rho$. Now, a few crucial questions remain:

- What is the form of $t[\rho(\mathbf{r})]$?
- What is the form of $v_{xc}[\rho(\mathbf{r})]$?
- How does one carry out practical calculations?

2.5.2. *Kohn–Sham equations*

Kohn and Sham (1965) introduced a scheme leading to the self-consistent method for calculating ground state electronic properties of a many-body system. There are several versions from the original formulation. We refer to the one by Grossu and Paravocicino (2000). The first step is to modify the total energy functional by specifying the density and add and subtract a non-interacting kinetic energy functional:

$$\rho(\mathbf{r}) = \sum_{i=\text{occ}} \psi_i^*(\mathbf{r})\psi_i(\mathbf{r}), \tag{2.56}$$

where ψ_i are fictitious, single-particle wave functions of the system,

$$T_0[\rho(\mathbf{r})] = \sum_i \left\langle \psi_i \left| \frac{-\hbar^2\nabla^2}{2m} \right| \psi_i \right\rangle, \tag{2.57}$$

and define a new exchange-correlation energy functional as:

$$E'_{xc}[\rho(\mathbf{r})] = E_{xc}[\rho(\mathbf{r})] + T[\rho(\mathbf{r})] - T_0[\rho(\mathbf{r})]. \tag{2.58}$$

The total energy functional is now modified,

$$F = \int T_0[\rho]d^3r + V_{\text{ext}}[\rho] + \frac{1}{2}\iint \frac{\rho(\mathbf{r})\rho(\mathbf{r}')}{|\mathbf{r}-\mathbf{r}'|}d^3rd^3r' + \int E'_{xc}[\rho]d^3r. \tag{2.59}$$

The single-particle Schrödinger equation is now modified to:

$$\left\{ \frac{-\hbar^2\nabla^2}{2m} + v_{\text{ext}}(r) + \int \frac{\rho(\mathbf{r}')}{|\mathbf{r}-\mathbf{r}'|}d^3r' + v'_{xc}[\rho(\mathbf{r})] - \lambda \right\} \psi_i(\mathbf{r}) = 0. \tag{2.60}$$

The self-consistent scheme is to solve Eqs. (2.56) and (2.60) self-consistently. We are still left with $v'_{xc}[\rho((\mathbf{r})]$ unspecified. Kohn and Sham (1965) proposed the LDA for the exchange-correlation energy functional.

2.5.2.1. *Local density approximation (LDA)*

In DFT, the exchange-correlation energy functional $E'_{\text{xc}}[\rho]$ in Eq. (2.58) is:

$$E'_{\text{xc}}[\rho] = \int v'_{\text{xc}}[\rho(\mathbf{r})]\rho(\mathbf{r})d^3r, \qquad (2.61)$$

with

$$v'_{\text{xc}}[\rho(\mathbf{r})] = \int \frac{g(\mathbf{r},\mathbf{r}')\rho(\mathbf{r}')}{|\mathbf{r}-\mathbf{r}'|}d^3r', \qquad (2.62)$$

where $g(\mathbf{r},\mathbf{r}')$ is the pair correlation function. It is a nonlocal function specifying the probability for a particle at $\mathbf{r}'$ when there is a particle at $\mathbf{r}$. In LDA, v'_{xc} is approximated as follows: Let

$$E_{\text{xc}}^{\text{LDA}}[\rho] = \int \varepsilon_{\text{xc}}[\rho(\mathbf{r})]\rho(\mathbf{r})d^3r, \qquad (2.63)$$

where $\varepsilon_{\text{xc}}[\rho(\mathbf{r})]$ is the exchange-correlation energy per particle. $\varepsilon_{\text{xc}}[\rho(\mathbf{r})]$ is determined locally by the density at $\mathbf{r}$. Thus $v'_{\text{xc}}[\rho(\mathbf{r})]$ can be expressed as:

$$v'_{\text{xc}}[\rho(\mathbf{r})] = \frac{\partial E_{\text{xc}}^{\text{LDA}}[\rho]}{\partial \rho} = \varepsilon_{\text{xc}}[\rho(\mathbf{r})] + \rho \left[\frac{\partial \varepsilon_{\text{xc}}[\rho]}{\partial \rho}\right]_{\rho=\rho(\mathbf{r})}. \qquad (2.64)$$

Equations (2.63) and (2.64) are the essence of LDA. $\varepsilon_{\text{xc}}[\rho(\mathbf{r})]$ is the sum of the exchange energy per particle, $\varepsilon_x[\rho(\mathbf{r})]$, and correlation energy per particle, $\varepsilon_c[\rho(\mathbf{r})]$. In terms of r_s — the average spacing between particles — Perdew and Zunger (1981) fit these energies to Monte Carlo results of Ceperley and Alder (1980) to obtain

$$\varepsilon_x[\rho(\mathbf{r})] = -\frac{0.4582}{r_s}, \qquad (2.65)$$

and

$$\varepsilon_c[\rho(\mathbf{r})] = \begin{cases} -\dfrac{0.1423}{1 + 1.0529\sqrt{r_s} + 0.3334 r_s} & r_s \geq 1 \\[2mm] -0.048 + 0.311\ln(r_s) - 0.0116 r_s + 0.002 r_s \ln(r_s) & r_s < 1. \end{cases} \qquad (2.66)$$

2.5.2.2. *Spin-polarized Kohn–Sham equations*

Half-metals are a class of FM materials. To study their magnetic properties, it is necessary to include the spin degree of freedom. In principle, we

should review DFT in spin-polarized form. The primary difference from the
formalism given in Section 2.5.2 is to generalize the energy density functional in terms of 2×2 matrix form, which was first given by von Barth
and Hedin (1972). Readers who are interested in this subject can read the
review by Zeller (2006). Here we only discuss the essentials of spin-polarized
Kohn–Sham equations (von Barth and Hedin, 1972) which are relevant to
half-metallic properties. Let α denote the $\uparrow$ spin and β denote the $\downarrow$ spin
states. We follow the formalism in LDA by defining

$$\rho_{\alpha\beta}(\mathbf{r}) = \sum_{i=occ} \psi_{\alpha,i}^*(\mathbf{r})\psi_{\beta,i}(\mathbf{r}). \tag{2.67}$$

For diagonal terms, i.e., $\alpha = \beta$, the two densities are denoted by either
$\rho_\alpha(\mathbf{r})$ or $\rho_\beta(\mathbf{r})$ and the Schrödinger equation is now in a coupled form,

$$\sum_\nu \left\{ \delta_{\mu\nu} \left(\frac{-\hbar^2}{2m}\nabla^2 + v_{\text{ext}}(\mathbf{r}) + \sum_\gamma \int \frac{\rho_{\gamma\gamma}(\mathbf{r})}{|\mathbf{r} - \mathbf{r}'|}d^3r' \right) + v'_{\text{xc}}[\rho(\mathbf{r})]_{\mu\nu} - \lambda \right\}$$
$$\psi_\nu = 0, \tag{2.68}$$

where μ, ν, and γ are either α or β. With the local-spin-density approximation (LSDA), the exchange-correlation functional $E_{\text{xc}}^{\text{LSDA}}$ is

$$E_{\text{xc}}^{\text{LSDA}}[\rho_{\alpha\beta}(\mathbf{r})] = \int \varepsilon_{\text{xc}}[\rho_\alpha(\mathbf{r}), \rho_\beta(\mathbf{r})](\rho_\alpha(\mathbf{r}) + \rho_\beta(\mathbf{r}))d^3r \tag{2.69}$$

$$v'_{\text{xc}}[\rho(\mathbf{r})]_{\alpha\beta} = \frac{\partial E_{\text{xc}}^{\text{LSDA}}[\rho]}{\partial \rho_{\alpha\beta}}. \tag{2.70}$$

Note the matrix forms in Eqs. (2.69) and (2.70). Additional information
about the magnetic moment at a point inside a unit cell and the magnetic
moment per unit cell can be obtained,

$$m(\mathbf{r}) = \rho_\alpha(\mathbf{r}) - \rho_\beta(\mathbf{r}), \tag{2.71}$$

and

$$M = \int m(\mathbf{r})d^3r. \tag{2.72}$$

Although simple, the LDA (or LSDA) results in a realistic description
of the atomic structure, elastic, and vibrational properties for a wide range
of systems. However, LDA is generally not accurate enough to describe the

energetics of chemical reactions (heats of reaction and activation energy barriers), leading to an overestimate of the binding energies of molecules and solids in particular. To improve this, a GGA exchange-correlation energy functional is often used. It has several versions, of which the most commonly used is the one by Perdew *et al.* (1992).

2.5.2.3. *Generalized gradient approximation (GGA)*

Basic idea To improve the calculated total energy of a physical system, Langreth and Mehl (1983), Becke (1988), and Perdew and Yue (1986) developed the GGA exchange-correlation functional. The essential idea is to generalize the LSDA exchange-correlation energy functional, $E_{xc}^{LSDA}[\rho_{\alpha\beta}(\mathbf{r})]$ given in Eq. (2.69) to

$$E_{xc}^{GGA}[\rho_\alpha(\mathbf{r}), \rho_\beta(\mathbf{r})] = \int f[\rho_\alpha(\mathbf{r}), \rho_\beta(\mathbf{r}), \nabla\rho_\alpha(\mathbf{r}), \nabla\rho_\beta(\mathbf{r})]d^3r, \quad (2.73)$$

where f is a functional of spin-polarized densities and their derivatives. There are several forms in the literature. The PBE exchange-correlation functional (Perdew, Burke and Ernzerhof, 1996, 1997) is a popular one which divides E_{xc}^{GGA} into the correlation part E_c^{GGA} and exchange part E_x^{GGA}.

Expressions for E_c^{GGA} and E_x^{GGA} For facilitating the numerical methods, Perdew *et al.* (1996) and Perdew *et al.* (1997) proposed analytic expressions for both energy functionals satisfying energetically significant conditions and having constant parameters. We start with E_c^{GGA},

$$E_c^{GGA}[\rho_\alpha(\mathbf{r}), \rho_\beta(\mathbf{r})] = \int \rho(\mathbf{r})\{\varepsilon_c[\rho_\alpha(\mathbf{r}), \rho_\beta(\mathbf{r})] + C(r_s, \rho_{\alpha-\beta}(\mathbf{r}), t)\}d^3r, \quad (2.74)$$

where ε_c is the one used in Eq. (2.66) and

$$\rho_{\alpha-\beta}(\mathbf{r}) = (\rho_\alpha(\mathbf{r}) - \rho_\beta(\mathbf{r}))/(\rho_\alpha(\mathbf{r}) + \rho_\beta(\mathbf{r})). \quad (2.75)$$

$t = |\nabla\rho(\mathbf{r})|/(2\phi k_{TF}\rho)$ with $k_{TF} = \sqrt{(4k_F)/(\pi a_B)}$ as the Thomas–Fermi screening wave number, where k_F is the Fermi momentum and a_B is the Bohr radius. Let $\phi = (1/2)[(1+\rho)^{2/3} + (1-\rho)^{2/3}]$, then

$$C(r_s, \rho_{\alpha-\beta}(\mathbf{r}), t) = \frac{e^2}{a_B}\gamma\phi^3 \ln\left\{1 + \frac{\beta(1+At^2)}{\gamma(1+At^2+A^2t^4)}t^2\right\}, \quad (2.76)$$

where $A = (\beta/\gamma)\{\exp(-\varepsilon_c/(e^2\gamma\phi^3/a_B)) - 1\}^{-1}$, $\gamma = (1 - \ln 2)/\pi^2$ and $\beta = 0.066725$. E_x^{GGA} is given as

$$E_x^{\mathrm{GGA}}[\rho_\alpha(\mathbf{r}), \rho_\beta(\mathbf{r})] = \int \rho(\mathbf{r})\{\varepsilon_x[\rho(\mathbf{r})]F_x(s)\}d^3r, \tag{2.77}$$

where $\varepsilon_x[\rho(\mathbf{r})] = -3e^2k_F/(4\pi)$, $F_x(s) = 1 + \eta - \frac{\eta}{1+0.21951s^2}$, $s = \frac{|\nabla\rho|}{2k_F\rho}$, and $\eta = 0.804$.

With the basic concept of DFT, we now discuss the essence of several popular methods for calculating the electronic properties of various materials, including HMs based on DFT.

2.5.3. *Methods of calculating electronic properties*

For the methods to be discussed, the electron exchange-correlation interactions are treated within either LSDA or GGA. The differences among these methods are in the treatments of ionic potentials, whether including core states or not, and the choice of basis functions.

2.5.3.1. *Linearized augmented plane wave (LAPW) method*

An important feature of this method is the inclusion of all electrons, both core and valence, in the calculation. Therefore, it is sometimes called an "all-electron" method. In principle, it is, to date, the most accurate method for calculating electronic properties. Because of the inclusion of all electrons, it is generally restricted to materials with a relatively small number (<50) of atoms per unit cell. The augmented plane wave (APW) method on which it is based was originally proposed by Slater (1951).

Basic idea In order to make the method tractable, Slater employed an approximation that the potential has spherical symmetry near an atom in a solid and is constant in the interstitial region: the so-called muffin-tin (MT) approximation. A schematic diagram of the MT potential is given in Fig. 2.38. The lattice is indicated by mesh. At each intersection, there is a lattice point and an atom. The MT potential around each atom is shown by a filled circle known as the MT sphere. Its radius is called the MT radius, R_m. The arrows mark the interstitial region.

With the MT potential, the wave function inside the MT sphere can be expanded in spherical harmonics, $Y_{lm}(\theta, \phi)$, where θ and ϕ are the standard angular variables. In the interstitial region the wave function can be

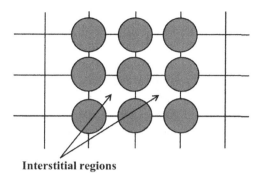

Interstitial regions

Fig. 2.38.   Schematic diagram of MT potential. Intersections of lattice lines represent lattice points of the crystal. The spherical potentials about each atom are shown as filled circles. The interstitial region is indicated by arrows.

expanded in plane waves $e^{i\mathbf{G}\cdot\mathbf{r}}$. An APW is then defined as

$$
\psi_{\mathrm{APW}}(\mathbf{r}) =
\begin{cases}
\displaystyle\sum_{l,m} A_{lm} R_l(r, E) Y_{lm}(\theta, \phi) & r \leq R_m \\
e^{i\mathbf{G}\cdot\mathbf{r}} & \text{interstitial,}
\end{cases}
\tag{2.78}
$$

where $R_l(r, E)$ is the solution of the radial equation with angular momentum l — the quantum number — at energy E and is a function of both r and E. A_{lm} is determined by matching the expansion inside the MT sphere and plane wave at R_m. The set of all such APWs is then used as the basis for the wave functions in the solid.

Secular equation   The variational principle is applied to the expansion of the wave function in terms of APWs. There are two points to note:

- The derivatives of the basis functions at $r = R_m$ are discontinuous.
- Because of the energy dependence of the $R_l(r, E)$ in the APW, the matrix elements of the secular equation are energy-dependent.

Therefore, many existing powerful diagonalization algorithms cannot be applied. A linearized formulation of the APW method was proposed by Andersen (1975), by expanding the wave function inside the MT on a complete set of states at a fixed energy E_l, and by Koelling and Arbman (1975), by making use of wave functions and their energy derivatives at different E_l. This method is now called the linearized augmented plane wave method.

Remarks   Since the basis functions of APWs are used, the matrix elements of the Coulomb and exchange-correlation potentials can be readily calculated. Over the years, a number of software implementations of the method have been developed. One popular and versatile implementation is the WIEN package.[1] For a detailed discussion of the LAPW method, the reader is referred to the excellent book by Singh and Nordström (2006).

2.5.3.2. *Korringa–Kohn–Rostoker (KKR) method*

Basic idea   The KKR method treats ionic potentials inside a solid as scattering centers. In a solid having one atom/unit cell, the lattice vectors $\mathbf{R}_n$ are defined with respect to a chosen origin. Here, n is in fact a set of three integers to specify the location of the n-th unit cell. At the origin of the lattice, a Wigner–Seitz (WS) cell is defined. The MT sphere S_m and the corresponding MT potentials are inside the WS cell. The Green's function is the response at a position $\mathbf{r}$ due to a unit source located at a point $\mathbf{r}'$ and is scattered by the MT potentials at other lattice sites.

In order to facilitate the calculations, non-overlapping MT potentials $\Sigma_n V_m(\mathbf{r} - \mathbf{R}_n)$ are used, where the summation runs over all lattice points and $V_m(\mathbf{r} - \mathbf{R}_n)$ is the MT potential at $\mathbf{R}_n$. The original KKR formulation applied to solids is based on the work of Dupree (1961) and Beeby (1967). It has been applied to half-metallic compounds by Podloucky *et al.* (1980) and Akai *et al.* (1985).

Green's function   The Schrödinger equation to be solved is

$$\left\{ \frac{-\hbar^2}{2m} \nabla^2 + \sum_n V_m(\mathbf{r} - \mathbf{R}_n) \right\} \psi(\mathbf{r}) = E\psi(\mathbf{r}). \tag{2.79}$$

The Green's function associated with Eq. (2.79) satisfies the following equation:

$$\{H_0 + E\}G(\mathbf{r} + \mathbf{R}_n, \mathbf{r}' + \mathbf{R}_{n'}; E) = -\delta(\mathbf{r} - \mathbf{r}')\delta_{nn'}(\mathbf{r}), \tag{2.80}$$

---

[1] http://www.wien2k.at

where H_o is $(-\hbar^2/2m)\nabla^2$. By constructing a functional Λ and setting $\partial\Lambda/\partial\psi^*(\mathbf{r}) = 0$, an integral equation is obtained, where

$$\Lambda = \int \psi^*(\mathbf{r}) \left\{ \sum_n V_m(\mathbf{r} - \mathbf{R}_n) \right\} \psi(\mathbf{r}) d^3r$$

$$+ \iint \psi^*(\mathbf{r}') \left\{ \sum_n V_m(\mathbf{r} - \mathbf{R}_n) \right\} G(\mathbf{r}, \mathbf{r}'; E)$$

$$\times \left\{ \sum_{n'} V_m(\mathbf{r}' - \mathbf{R}_{n'}) \right\} \psi(\mathbf{r}') d^3r d^3r' \qquad (2.81)$$

and

$$\psi(\mathbf{r}) = \int G(\mathbf{r}, \mathbf{r}'; E) \left\{ \sum_{n'} V_m(\mathbf{r}' - \mathbf{R}_{n'}) \right\} \psi(\mathbf{r}') d^3r'. \qquad (2.82)$$

By changing the variable $\mathbf{r}'$ and applying the Bloch theorem,

$$\psi_{\mathbf{k}}(\mathbf{r} + \mathbf{R}_n) = e^{i\mathbf{k}\cdot\mathbf{R}_n}\psi_{\mathbf{k}}(\mathbf{r}), \qquad (2.83)$$

where $\mathbf{k}$ is a vector inside the first Brillouin zone (BZ), Green's function is now $\mathbf{k}$-dependent and is given by:

$$G_{\mathbf{k}}(\mathbf{r} - \mathbf{r}'; E) = \sum_n e^{-i\mathbf{k}\cdot\mathbf{R}_n} G_{\mathbf{k}}(\mathbf{r} - \mathbf{r}' - \mathbf{R}_n; E). \qquad (2.84)$$

In Eq.(2.84), we assume that space is homogeneous: Green's function depends on relative displacements of the source and point of observation. The boundary condition satisfied by the wave function $\psi(\mathbf{r})$ is obtained by multiplying Eq. (2.79) by G and Eq. (2.80) by $\psi(\mathbf{r})$, then subtracting the two resulting equations, and finally setting the surface integral term to be zero:

$$\int_S \left\{ G_{\mathbf{k}}(\mathbf{r} - \mathbf{r}'; E) \frac{\partial\psi(\mathbf{r}')}{\partial\mathbf{r}'} - \psi(\mathbf{r}') \frac{\partial G_{\mathbf{k}}(\mathbf{r} - \mathbf{r}'; E)}{\partial\mathbf{r}'} \right\} dS = 0. \qquad (2.85)$$

To apply Eq. (2.85), an expression for G is necessary. Eq. (2.80) can be easily solved for a single site,

$$G(\mathbf{r} - \mathbf{r}'; E) = \frac{1}{E - H_0}. \qquad (2.86)$$

By using the eigenfunctions of H_o — the plane waves — the spectral representation of Green's function with summation over BZ can be obtained. After carrying out the sum, the real-space representation of Green's function is given as:

$$G(\mathbf{r} - \mathbf{r}'; E) = -\frac{e^{i\kappa|\mathbf{r}-\mathbf{r}'|}}{4\pi|\mathbf{r} - \mathbf{r}'|}, \qquad (2.87)$$

where κ is $\sqrt{E}$ if E is positive, and is $i\sqrt{|E|}$ if E is negative. Equation (2.87) can be expanded in spherical harmonics, the so-called Neumann expansion,

$$-\frac{e^{i\kappa|\mathbf{r}-\mathbf{r}'|}}{4\pi|\mathbf{r} - \mathbf{r}'|} = \kappa \sum_{l,m} j_l(\kappa r) Y_{lm}(\theta, \phi)\{n_l(\kappa r') - ij_l(\kappa r')\} Y_{lm}^*(\theta', \phi') \quad r < r',$$

$$(2.88)$$

where (r, θ, ϕ) are the usual spherical coordinates. j_l is the spherical Bessel function of the first kind and n_l is the spherical Bessel function of the second kind. For $r > r'$, one exchanges r and r' at the right-hand side of Eq. (2.88).

Structure factor   Since the KKR method is based on the idea that the electron wave function at a point $\mathbf{r}$ is a consequence of interferences of scattered waves from the potential centers, the important quantity is the so-called structure factor which specifies phase relations of the potential centers. For one atom per unit cell, Eq. (2.84) provides the basic form of the structure factor. By substituting Eq. (2.87) into Eq. (2.84), one gets

$$G_{\mathbf{k}}(\mathbf{r} - \mathbf{r}'; E) = -\frac{\cos(\kappa|\mathbf{r} - \mathbf{r}'|)}{4\pi|\mathbf{r} - \mathbf{r}'|}$$

$$+ \sum_{l,m} \Gamma_{lm}(k, E) j_l(\kappa|\mathbf{r} - \mathbf{r}'|) Y_{lm}(\theta_{\mathbf{r}-\mathbf{r}'}, \phi_{\mathbf{r}-\mathbf{r}'}), \qquad (2.89)$$

where $\theta_{\mathbf{r}-\mathbf{r}'}$ is the polar angle and $\phi_{\mathbf{r}-\mathbf{r}'}$ is the azimuthal angle of the vector $\mathbf{r} - \mathbf{r}'$, respectively. $\Gamma_{lm}(k, E)$ is the structure factor and is given by

$$\Gamma_{lm}(k, E) = \kappa \sum_{n} e^{-i\mathbf{k}\cdot\mathbf{R}_n}\{n_l(\kappa R_n) - ij_l(\kappa R_n)\} - i\kappa \frac{\delta_{l0}\delta_{m0}}{\sqrt{4\pi}}. \qquad (2.90)$$

For the case of many atoms per unit cell, Eq. (2.90) can be easily modified to accommodate the situation. The most recent improvement including the coherent potential method is given by Akai and Dederichs (1993).

Eigenvalue problem   After obtaining Green's function, one should be able to find the eigenvalues by using a trial wave function and Eq. (2.84).

A possible trial wave function is

$$\psi(\mathbf{k}, \mathbf{r}) = \sum_{l,m} c_{lm}(\mathbf{k}) R_l(E, r) Y_{lm}(\theta, \phi), \qquad (2.91)$$

where R_l is the regular solution inside the MT sphere. Outside the MT sphere, the potential is expressed in terms of phase shifts which are functions of energy E. From Eq. (2.84), a set of linear coupled equations for $c_{lm}(\mathbf{k})$ can be set up. The corresponding secular equation gives eigenvalues at each $\mathbf{k}$-point.

2.5.3.3. *Pseudopotential method*

This is one of the more popular methods for calculating electronic properties in condensed matter. There are many well-documented algorithms available either on the market, such as VASP,[2] or free, such as ABINIT,[3] QuantumEspresso,[4] and CP2K.[5] In this section, we discuss the basic idea of pseudopotentials and comment on recent developments in the area.

Basic idea The simplest way to express a valence electron in a material is to use a plane wave $|\mathbf{k}\rangle = \frac{1}{\sqrt{\Omega}} e^{i\mathbf{k}\cdot\mathbf{r}}$. The amplitude of a plane wave is a constant and Ω is the unit-cell volume. In reality, the wave function of a valence state must be orthogonal to core states. Herring (1940) introduced the orthogonalized plane wave, OPW, to take into account this orthogonalization constraint.

$$|\text{OPW}\rangle = |\mathbf{k}\rangle - \sum_c |c\rangle\langle c|\mathbf{k}\rangle, \qquad (2.92)$$

where $|c\rangle$ is the core wave function in Bloch form. For treating electronic states in a real solid with periodicity, the description of a simple plane wave is not sufficient. A Bloch form of an $|\text{OPW}\rangle$ should be more appropriate. Then, the orthogonal processes become a tremendous burden. Phillips and Kleinman (1959) reformulated the orthogonalization process into a form of potential — the pseudopotential. The form of a pseudopotential is

$$V_{\text{ps}}(\mathbf{r}) = V(\mathbf{r}) + \sum_c (E - E_c)|c\rangle\langle c|, \qquad (2.93)$$

---

[2]http://www.vasp.at
[3]http://www.abinit.org
[4]http://www.quantum-espresso.org
[5]http://www.cp2k.org

where $V(\mathbf{r})$ is the crystal potential. E is the eigenvalue for a valence state. E_c is the energy of a core state. Some features of pseudopotentials are:

- The second term is repulsive because $E > E_c$. $|V_{ps}(\mathbf{r})|$ is therefore weaker than $|V(\mathbf{r})|$ in the core region.
- $V_{ps}(\mathbf{r})$ is energy-dependent.
- $V_{ps}(\mathbf{r})$ is nonlocal because of the presence of projection operators.

Key developments The semi-nonlocal form — only spherical harmonics Y_{lm} are used in the projection operators — was used by Lee and Falicov (1968) to calculate the band structure of potassium, and by Fong and Cohen (1970) to calculate the band structure of copper. Because of the nonseparable form of the semi-nonlocal operators, however, a large number of off-diagonal matrix elements needs to be computed and stored. Kleinman and Bylander (1982) proposed an approximate separable form for nonlocal operators to reduce computation and storage. In the late 1970s and early 1980s, implementation of first-principles norm-conserving pseudopotentials in plane-wave-based DFT was developed (Hamann *et al.*, 1979). Since the pseudopotential in the core region is not unique (Austin *et al.*, 1962), and it is appealing to use a plane-wave basis for its flexibility and efficiency, Vanderbilt (1990) proposed ultrasoft pseudopotentials for elements with tightly bound, e.g., O $2p$ and Ni $3d$, orbitals. This development enables one to treat TMs efficiently with pseudopotentials in a plane-wave basis. Recent development by Blöchl (1994) and Kresse and Joubert (1999) of the projector augmented wave (PAW) method has shown that the method can be as accurate as conventional all-electron methods while retaining the advantages of a plane-wave basis. The PAW formulation is now available in widely used codes such as VASP and ABINIT.

2.5.3.4. *LDA+U*

There are some concerns about electron-correlation effects on electronic properties of Heusler alloys and oxides. Here we briefly discuss the essence of electron-correlation treated in these two classes of HMs.

The most important electron-correlation effect due to localized d-electrons in TM elements, such as Co, Cr, Fe, and Mn involved in Heusler alloys and oxides, is the Coulomb repulsion experienced by localized electrons. The LDA+U (including LSDA+U) scheme is designed to model localized states when on-site Coulomb interactions become important. Since most modern DFT implementations such as VASP and WIEN have included the LDA+U scheme, we explain the basic ideas and comment on cautions one should take when using the scheme.

Basic idea The concept of the U term characterizing the Coulomb interaction at an atomic site in a solid was introduced by Hubbard (1963). The starting point is a many-electron Hamiltonian,

$$H = \sum_i \left\{ -\frac{\hbar^2}{2m} \nabla_i^2 + V_{\text{ext}}(\mathbf{r}_i) \right\} + \sum_i \sum_{j \neq i} \frac{e^2}{|\mathbf{r}_i - \mathbf{r}_j|}, \qquad (2.94)$$

where i and j are dummy variables to enumerate the electrons. The Coulomb interaction $\frac{e^2}{|\mathbf{r}_i - \mathbf{r}_j|}$ can be replaced by screened electron–electron interaction V_{e-e}. By using a set of localized functions, say, the Wannier functions, the Hamiltonian is second quantized. Let the set of Wannier functions be $\{w(\mathbf{r}_i)\}$, then the matrix element of the single-particle kinetic energy and the external potential, $\langle w(\mathbf{r}_i)| -\frac{\hbar^2}{2m} \nabla_i^2 + V_{\text{ext}}(\mathbf{r}_i)|w(\mathbf{r}_i)\rangle$, defines the on-site energy ε_i. The matrix element, $\langle w(\mathbf{r}_i)| -\frac{\hbar^2}{2m} \nabla_i^2 |w(\mathbf{r}_j)\rangle$, describes the hopping integral t_{ij} of an electron hopping from site i to site j. The matrix element of the two-particle operator $V_{e-e}(|\mathbf{r}_i - \mathbf{r}_j|)$ gives the strength of Coulomb repulsion $\langle w(\mathbf{r}_i)w(\mathbf{r}_j)|V_{e-e}(|\mathbf{r}_i - \mathbf{r}_j|)|w(\mathbf{r}_i)w(\mathbf{r}_j)\rangle$. This term is approximated as $U n_{i\uparrow} n_{j\downarrow}$, where $n_{i\sigma} = a_{i\sigma}^\dagger a_{i\sigma}$ and $a_{i\sigma}^\dagger$ and $a_{i\sigma}$ are the creation and annihilation operators at the i-th site with spin σ. If i and j are equal, it defines the on-site U term. The so-called Hubbard model Hamiltonian H_H is,

$$H_H = \sum_{i,\sigma} \varepsilon_i a_{i\sigma}^\dagger a_{i\sigma} + \sum_{i,\sigma} \sum_{j \neq i, \sigma'} t_{i\sigma, j\sigma'} (a_{i\sigma}^\dagger a_{j\sigma'} + H.c.) + \sum_i U n_{i\uparrow} n_{i\downarrow},$$
$$(2.95)$$

where $H.c.$ is the Hermitian adjoint of $a_{i\sigma}^\dagger a_{j\sigma'}$.

Equation (2.95) is a many-body expression. To implement U in DFT formalism, the general procedure is to add the U term to either LDA or GGA total energy. The total energy of a physical system can then be written as (Hubbard, 1963):

$$E^{\text{Tot}}\left[\rho_\sigma, \rho_{\sigma'}; o_{i\sigma}^{nlm}, o_{i\sigma'}^{n'l'm'}\right] = E^{\text{LDA}}[\rho_\sigma, \rho_{\sigma'}] + E^U\left[o_{i\sigma}^{nlm}, o_{i\sigma'}^{n'l'm'}\right]$$
$$- E^{\text{dc}}\left[o_{i\sigma}^{nlm}, o_{i\sigma'}^{n'l'm'}\right], \qquad (2.96)$$

where E^{LDA} is the total energy of LDA or GGA functional, E^U is the total energy derived from the U term, and E^{dc} is the double counting total energy to correct the overlap between E^{LDA} and E^U energies. $o_{i\sigma}^{nlm}$ is the trace of the local density matrix $\hat{O}_{i\sigma,j\sigma'}^{nlm,n'l'm'}$, where i is the site index, nlm is the product of the principal, angular momentum, and magnetic quantum numbers.

Implementation of Hubbard U To implement the last two terms of Eq. (2.96), it is important to realize that:

- E^U should be calculated within the atomic region where atomic characters of electronic states can be identified. Assuming it is possible to identify atomic orbitals as $|inl; m\sigma\rangle$ within an atomic sphere then only $|m\sigma\rangle$ is of interest. Thus the notation of $|inl; m\sigma\rangle$ can be simplified as $|m\rangle$.

$$
E^U\left[o_{i\sigma}^{nlm}, o_{i\sigma'}^{n'l'm'}\right]
$$

$$
= \frac{1}{2} \sum_{\{m\},\sigma} \left\{ \begin{array}{l} \langle m, m''|V_{e-e}|m', m'''\rangle \hat{O}_{\sigma,\sigma}^{m,m'} \hat{O}_{-\sigma,-\sigma}^{m'',m'''} \\ + [\langle m, m''|V_{e-e}|m', m'''\rangle \\ - \langle m, m''|V_{e-e}|m''', m'\rangle] \hat{O}_{\sigma,\sigma}^{m,m'} \hat{O}_{\sigma,\sigma}^{m'',m'''} \end{array} \right\}.
$$

$$
\tag{2.97}
$$

If the density matrix is diagonal then $\hat{O}_{\sigma,\sigma}^{m,m'} = \hat{O}_{\sigma,\sigma}^{m,m} \delta_{m,m'}$.

- If there is no orbital polarization, Eq. (2.96) should be reduced to E^{LSDA} and E^{dc} can be expressed in terms of U and J — the exchange integral (Liechtenstein *et al.*, 1995).

$$
E^{dc}\left[o_{i\sigma}^{nlm}, o_{i\sigma'}^{n'l'm'}\right] = \frac{1}{2}U\{o(o-1)\} - \frac{1}{2}J\{o_\uparrow(o_\uparrow - 1) + o_\downarrow(o_\downarrow - 1)\},
$$

$$
\tag{2.98}
$$

where $o = o_\uparrow + o_\downarrow$ and o_σ are trace of $\hat{O}_{\sigma,\sigma}^{m,m'}$. In practice, an effective single-particle potential is deduced from the above equations and added to V^{LSDA},

$$
V_\sigma^{m,m'} = \sum_{m'',m'''} \left\{ \begin{array}{l} \langle m, m''|V_{e-e}|m', m'''\rangle \hat{O}_{-\sigma,-\sigma}^{m'',m'''} \\ + [\langle m, m''|V_{e-e}|m', m'''\rangle \\ - \langle m, m''|V_{e-e}|m''', m'\rangle] \hat{O}_{\sigma,\sigma}^{m'',m'''} \end{array} \right\}
$$

$$
- U\left(o - \frac{1}{2}\right) + J\left(o_\sigma - \frac{1}{2}\right).
$$

$$
\tag{2.99}
$$

- The matrix elements of V_{e-e} are expressed in terms of Slater integrals $F^k(nl, nl')$ with $0 \leq k \leq 2l$.

$$\langle m, m'' | V_{e-e} | m', m''' \rangle = \sum_k a_k(m, m', m'', m''') F^k,$$

$$a_k(m, m', m'', m''') = \frac{4\pi}{2k+1} \sum_{q=-k}^{k} \langle lm | Y_{kq} | lm' \rangle \langle lm'' | Y_{kq}^* | lm''' \rangle,$$

(2.100)

where Y_{kq} is a spherical harmonics.

Typically, U value is approximately 20 eV in a free TM element and is on the order of a few eV in a solid.

2.5.4. *Methods of calculating Curie temperature T_C*

To calculate T_C, there are a few methods based on DFT. The foundation of these methods is the "magnetic force theorem" first proved by Mackintosh and Andersen (1980) and Heine (1980). It was then extended by Oswald *et al.* (1985) for systems with magnetic impurities. The theorem is frequently used to compute exchange interaction parameters and adiabatic spin-wave spectra of ferromagnets. The interest of this approach is the energy difference of different spin orientations at the atoms, which can be determined by the change of the one-particle energy. It allows the calculation of the change of the total energy to be carried out non-self-consistently. Thus, the theorem provides a way to reduce the computational effort. Liechtenstein *et al.* (1987) used the theorem for treating spin-wave spectra of FM systems within mean field theory (MFT). Bruno (2003) and his collaborators extended the applications to varieties of FM metallic systems and improved the MFT scheme proposed by Liechtenstein *et al.*

2.5.4.1. *Determination of the dominant excitation*

Pajda *et al.* (2001) determine the dominant excitation in FM materials using a first-principles real-space approach. In transition metals, such as Fe, Co, and Ni — the so-called itinerant ferromagnets — there are possibly two kinds of magnetic excitations: the Stoner excitation and the spin-wave excitation. The former involves interband transitions from the occupied ↑ spin band to an unoccupied ↑ spin band. They contribute to the longitudinal fluctuation of the magnetization. The latter is a collective excitation of spin moments of atoms. These collective excitations are called spin waves or magnons and contribute to the transverse fluctuation of the magnetization. The lower energies and corresponding large DOS of the magnon excitations cause them to be the dominant excitations at low temperature. Therefore, Pajda *et al.* considered only magnon excitations to determine T_C. This type of excitations is also adopted by others (Bouzerar *et al.*, 2006). For HMs, it is possible to use the same approach to determine T_C because the metallic channel contributes to the FM property.

2.5.4.2. *Basic idea*

The basic idea is to first map the interested spin system onto an effective Heisenberg Hamiltonian (Pajda *et al.*, 2001),

$$H_{\text{eff}} = -\sum_{i<j} J_{ij} \mathbf{S}_i \cdot \mathbf{S}_j, \tag{2.101}$$

where J_{ij} is the pairwise exchange interaction energy, $\mathbf{S}_i$ is the spin moment at site-i. If J_{ij} is obtained, then the magnon energy $\hbar\omega(\mathbf{q})$), where $\mathbf{q}$ is the momentum of the magnon, can be calculated,

$$\hbar\omega(\mathbf{q}) = \frac{4\mu_B}{m} \sum_j J_{0j}(1 - e^{i\mathbf{q}\cdot\mathbf{R}_{oj}}), \tag{2.102}$$

where m is the magnetic moment per atom, and $\mathbf{R}_{0j} = \mathbf{R}_0 - \mathbf{R}_j$ is the lattice vector at the j-lattice point from some chosen reference $\mathbf{R}_0$.

To determine T_C within the mean field approximation (MFA) (Bouzerar *et al.*, 2006), one then lets the energy of the magnetic system expressed in

$$E = g\mu_B \left\langle \sum_i \mathbf{h}_i \cdot \mathbf{S}_i \right\rangle = -\left\langle \sum_i \mathbf{h}_i \cdot \mathbf{m}_i \right\rangle = -\langle \mathbf{h} \cdot \mathbf{m} \rangle, \tag{2.103}$$

where $\mathbf{h}_i = \frac{2}{g\mu_B} \sum_{j\neq i} J_{ij} \mathbf{S}_j$ — the mean magnetic field at site i produced by magnetic moments from other sites, and g is the so-called g factor — the gyromagnetic ratio. For electron without spin–orbit coupling $g = 2$. The magnetic moment $\mathbf{m}_i$ at site-i is $-g\mu_B \mathbf{S}_i$. Or, the energy is expressed in terms of an effective mean field $\mathbf{h} = \lambda\mathbf{m}$, λ is the mean field parameter, and $\mathbf{m} = \sum_i \mathbf{m}_i$. The thermal average of $\mathbf{S}_j$ is expressed as:

$$\langle \mathbf{S}_j \rangle = \frac{Tr\left[\mathbf{S}_j e^{-i\frac{E}{k_B T}}\right]}{Tr\left[e^{-i\frac{E}{k_B T}}\right]}, \tag{2.104}$$

where k_B is the Boltzmann constant. By solving Eqs. (2.103) and (2.104) self-consistently, T_C^{MFA} is given by

$$k_B T_C^{\text{MFA}} = \frac{\lambda}{3} \langle \mathbf{m}^2 \rangle, \qquad (2.105)$$

where $\langle \mathbf{m}^2 \rangle$ is calculated at $T = 0\,\text{K}$ and λ is assumed to be independent of temperature.

A more accurate way is to use Green's function method. Callen (1963) used temperature-dependent Green's function and decoupled the hierarchy of the equation of motion by a physical intuitively appealing scheme. The magnetization of general spin is simply expressed as the z-component of the averaged spin operator for each site $\langle S^z \rangle$, and relates to the Bose–Einstein (B–E) distribution function Φ characterizing spin-wave excitations.

$$\langle S^z \rangle = \frac{(S - \Phi)(1 + \Phi)^{2S+1} + (1 + S + \Phi)\Phi^{2S+1}}{(1 + \Phi)^{2S+1} - \Phi^{2S+1}}, \qquad (2.106)$$

where

$$\Phi = \frac{1}{N} \sum_{\mathbf{q}} \frac{1}{e^{\frac{\hbar \omega(\mathbf{q})}{k_B T}} - 1}, \qquad (2.107)$$

where N is the number of $\mathbf{q}$ points. Bouzerar *et al.* (2006) used the random phase approximation (RPA) to decouple Green's function and derived similar expression for the magnetization. By expanding the B–E distribution, T_C is given by:

$$\frac{1}{k_B T_C^{\text{RPA}}} = \frac{6\mu_B}{M} \lim_{z \to 0} \frac{1}{N} \sum_{\mathbf{q}} \frac{1}{z - \hbar \omega(\mathbf{q})}. \qquad (2.108)$$

The term $\frac{1}{N} \sum_{\mathbf{q}} \frac{1}{z - \hbar \omega(\mathbf{q})}$ is in fact the magnon Green's function.

2.5.4.3. *Comments on practical calculations*

In practice, calculations of T_C involve several demanding steps. The first is the determination of J_{ij}; next, is the summation over $\mathbf{q}$, in Eq. (2.108).

Determination of J_{ij} There are two ways to determine J_{ij} and the excitation energy. One is the real-space scheme and the other is the $\mathbf{q}$-space or "frozen magnon" scheme.

Real-space approach In Eq. (2.101), for each i one calculates J_{ij} by considering the j-th shell of atoms. First-principles methods can be used to determine J_{ij} by calculating the change in energy when spin moments at the i-th and j-th sites are rotated by equal and opposite amounts with respect to the ground state (all spins aligned), and subtracting the changes in energy for the isolated rotations at the two sites, respectively. The tight-binding linearized MT orbital (TB-LMTO) method has been used by Pajda *et al.* (2001) with j up to 172nd shell for fcc structure and 195th shell for bcc lattice. In order to get converged $\omega(\mathbf{q})$, about 200 terms of J_{ij} are needed.

Momentum-space approach In this approach, one specifies the spin moments on the neighboring atoms to have a spiral configuration and carries out first-principles calculations to get the $\hbar\omega$ as a function of $\mathbf{q}$ relations for the configuration. The exchange interaction J_{ij} is obtained by inverse Fourier transform — an important contribution by Sandratskii (1998) — so that a large supercell is not necessary. The main relevant physics is illustrated in Fig. 2.39 using a one-dimensional model, where each arrow makes a polar angle θ with the z-axis and an azimuth angle φ with respect to the x-axis. The lattice vector is $\mathbf{R}_n$ where n represents the n-th lattice site. The moment at the n-th lattice site for a spiral structure with momentum $\mathbf{q}$ is given by:

$$\mathbf{M}_n = |\mathbf{M}_n|\{\cos(\mathbf{q} \cdot \mathbf{R}_n + \varphi_n)\sin\theta_n\hat{\mathbf{x}} + \sin(\mathbf{q} \cdot \mathbf{R}_n$$
$$+ \varphi_n)\sin\theta_n\hat{\mathbf{y}} + \cos\theta_n\hat{\mathbf{z}}\}, \tag{2.109}$$

where $|\mathbf{M}_n|$ is the magnitude of the spin moment at the n-th lattice site. It is easy to see that when there is a lattice translation $\mathbf{R}_n$, correspondingly there is a spin moment rotation by an angle of $-\mathbf{q} \cdot \mathbf{R}_n$ which leaves the

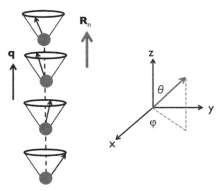

Fig. 2.39.   One-dimensional model showing the relative phase of the spin-spiral structure (Sandratskii, 1998).

invariance of the spiral structure. By analogy to notations used in the space group, one can define $(\{\mathbf{q} \cdot \mathbf{R}_n | e | \mathbf{R}_n\})$ as a generalized translation of the spiral structure, where e indicates that it does not have any real-space coordinate transformation, such as a rotation or a reflection and commutes with the Hamiltonian of the spiral structure. Because of e, it has no effect on the periodic part of the Bloch wave function — just a unit operator. The effect of generalized translation on a spinor wave function with electron momentum $\mathbf{k}$ is

$$P_{\mathbf{k}}(\{\mathbf{q} \cdot \mathbf{R}_n | e | \mathbf{R}_n\})\psi_{\mathbf{k}} = U(\alpha_S)e^{-i\mathbf{k}\cdot\mathbf{R}_n}\psi_{\mathbf{k}}, \tag{2.110}$$

where $P_{\mathbf{k}}$ is a unitary operator associated with the generalized translation operator. The active point of view is adopted here for the coordinate transformation on the wave function, where the transformation is:

$$U(\alpha_S) = \begin{pmatrix} e^{-\frac{i}{2}\mathbf{q}\cdot\mathbf{R}_n} & 0 \\ 0 & e^{\frac{i}{2}\mathbf{q}\cdot\mathbf{R}_n} \end{pmatrix}. \tag{2.111}$$

Equation (2.110) is the generalized Bloch theorem. For each $\mathbf{k}$-point, the eigenvalue $\varepsilon_n(\mathbf{k})$ of the nonmagnetic case is changed to $\varepsilon_n(\mathbf{k} - \frac{\sigma\mathbf{q}}{2})$, where σ is "+" or "−", the spin projection along the z-axis. $E(\mathbf{q})$ can be obtained by summing all occupied-state energies and subtracting the energy of the reference state. From $E(\mathbf{q})$ one can get $J(\mathbf{q})$, then J_{ij} is obtained by inverse Fourier transform.

Comparing the two schemes, the $\mathbf{k}$-space scheme is simpler owing to the generalized Bloch theorem. However, the calculations are still demanding in order to obtain accurate $\omega(\mathbf{q})$.

Summation over q To calculate T_C, there is still the $\mathbf{q}$-space sum to be carried out independent of MFA and RPA. It has to be done with extreme care about the $\mathbf{q} = 0$ point. A large number of $\mathbf{q}$-points should be used in order to obtain converged results.

Chapter 3

Heusler Alloys

3.1. Introduction

Heusler alloys are ternary intermetallic compounds. It was Heusler who first synthesized Cu_2MnAl in 1903 (Heusler, 1903). Excellent reviews have been given by Westerholt et al. (2005) and Hirohata et al. (2006). In 1971, Webster reported the chemical and magnetic structure of Co_2MnX alloys, with X = Si, Ge, and Sn. In 1983, de Groot et al. predicted that NiMnSb, now classified as a "half"-Heusler (HH) alloy, should exhibit half-metallic properties. Since then, more than 1,000 Heusler alloys have been synthesized. A typical chemical formula, X_2YZ, is adopted for the class of "full"-Heusler (FH) alloys that are predicted to have half-metallic properties, where X is a TM element, e.g., Co, Fe, Ni, or Pt; Y is another TM element, such as Cr, Mn, or Ti; and Z is an atom belonging to Group III, IV, or V, e.g., Al, Ge, Si, or Sb.

The recent development of spintronics has generated much interest in Heusler alloys. For spintronic applications, two criteria are essential:

(1) Effective injection of spin-polarized carriers into semiconductors.
(2) Sufficiently high T_C for devices to operate at RT.

The Heusler alloys are expected to meet both of these criteria because their structures are compatible with many semiconductors in both elemental and compound forms. The T_Cs of Heusler alloys are well above RT. However, only NiMnSb (Hanssen et al., 1990; Hordequin et al., 1996) has been found to exhibit half-metallic properties within 1/100 of an electron accuracy per formula unit experimentally. A number of these alloys have been predicted to be half-metallic theoretically (Wurmehl et al., 2005). These are excellent candidates for spintronic materials.

In this chapter, we focus on NiMnSb, Co_2FeSi, and a number of other Heusler alloys predicted by theory to exhibit half-metallic properties. There are two categories (Galanakis, 2002b) of Heusler alloys: the HH alloys, such

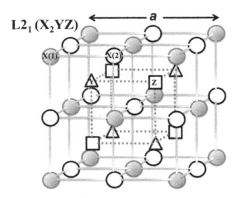

Fig. 3.1. $L2_1$ crystal structure with chemical formula, X_2YZ. $X(1)$ sites are denoted by filled circles, $X(2)$ by open circles, Y by open triangles, and Z by open squares. This structure corresponds to an FH alloy. The outermost cube edge has length a. If those sites with open circles $(X(2))$ are unoccupied, the structure corresponds to an HH alloy and is denoted by $C1_b$ (Galanakis *et al.*, 2002a).

as NiMnSb and PtMnSb, and the FH alloys, such as Co_2CrAl, Co_2FeAl, Co_2FeGe, Co_2MnSi, and Co_2FeSi. The structure of the first category $C1_b$ (Fig. 3.1) differs from the second category $L2_1$ in the occupation of TM sites in the unit cell. The $C1_b$ structure corresponds to the $L2_1$ structure with the $(X(2))$ sublattice unoccupied (Mancoff *et al.*, 1999; Galanakis *et al.*, 2002a).

In Section 3.2, we list the half-metallic HH and FH alloys determined by experiment or theory. The growth of these Heusler alloys will be discussed in Section 3.3, including both bulk and thin-film forms. Structural properties of FH and HH alloys will be discussed in Section 3.4.1. Bulk and thin-film properties will be described in Section 3.5. Finally, in Section 3.6 we discuss one specific HH alloy, NiMnSb, followed by the much-studied FH alloy, Co_2MnSi, and the FH alloy with the highest T_C, Co_2FeSi.

3.2. Half-Heusler and Full-Heusler Alloys

A great many Heusler alloys have been studied. However, no HM has been confirmed experimentally except NiMnSb (Hanssen *et al.*, 1990; Hordequin *et al.*, 1996). They have been mostly predicted by first-principles calculations. Tables 3.1 and 3.2 list HH and FH alloys, respectively, known to exhibit half-metallic properties experimentally or predicted theoretically to have integer magnetic moment/unit cell and DOS showing metallic behavior in one spin channel and insulating behavior in the other. We also give

Table 3.1. The magnetic moment/unit cell and the measured T_C of HF alloys.

Compound	Magnetic moment (μ_B/unit cell)		T_C (K)	Reference
	Exp.	Theory	Exp.	
CoMnSb	4.0		490	Webster and Ziebeck (1988)
		2.949		Galanakis (2005)
	4.2		478	Otto *et al.* (1989)
FeMnSb		1.930		Galanakis (2005)
NiMnSb	4.0			Hanssen *et al.* (1990)
	4.01 ± 0.02			Hordequin *et al.* (1996)
	4.02			Ritchie *et al.* (2003)
		4.0		de Groot *et al.* (1983)
	3.6			Clowes *et al.* (2004)
	3.85		730	van Engen *et al.* (1983)
	3.9 ± 0.2			Turban *et al.* (2002)
		3.991		Galanakis (2005)
		4.0		Halilov and Kulatov (1991)
		4.0		Block *et al.* (2004)
	4.2		728	Otto *et al.* (1989)
PdMnSb	3.95		500	van Engen *et al.* (1983)
		4.05		de Groot *et al.* (1983)
		4.062		Galanakis (2005)
PtMnSb	3.97		582	van Engen *et al.* (1983)
	4.14		572	Webster and Ziebeck (1988)
		4.0		de Groot *et al.* (1983)
	3.96		572	Otto *et al.* (1989)
		3.997		Galanakis (2005)
		4.003		Halilov and Kulatov (1991)
PtMnSn	3.42		330	van Engen *et al.* (1983)
		3.60		de Groot *et al.* (1983)
		3.51		Halilov and Kulatov (1991)
	3.5		330	Otto *et al.* (1989)

the available measured and predicted magnetic moment/unit cell, M, for each compound. M should, in principle, be an integer for a half-metallic compound, especially determined by theory.

Table 3.3 lists a few other FH alloys besides Co-related compounds predicted to show half-metallic properties.

3.3. Methods of Growing Heusler Alloys

The methods of growing Heusler alloys can be generally classified into two categories: the growth of bulk samples and the growth of thin-film samples. Given the anticipation that spintronic device applications will require

Table 3.2.   The magnetic moment/unit cell and the measured T_C of FH alloys.

| Compound | Magnetic moment (μ_B/unit cell) | | T_C (K) | Reference |
	Exp.	Theory	Exp.	
Co_2CrAl		4.811		Antonov et al. (2005)
		2.999		Galanakis (2005)
	0.53–0.86		310–330	Hirohata et al. (2005)
		3.0		Block et al. (2004)
Co_2FeAl	5.29 (4K)			Elmers et al. (2004)
	4.9			Kelekar and Clemens (2004)
	4.6–4.8			Hirohata et al. (2005)
		4.996		Galanakis (2005)
Co_2FeSi	5.492 (RT),5.907(10.2K)		>980	Niculescu et al. (1977)
	5.492 (RT),5.97(5.0K)	6.000	1100	Wurmehl et al. (2005)
	5.91			Hashimoto et al. (2005)
Co_2MnGe	5.11			Ambrose et al. (2000)
	4.84		905	Brown et al. (2000)
		5.00		Fuji et al. (1990)
		5.012		Galanakis (2005)
	4.93			Miyamoto et al. (2004)
	5.0			Picozzi et al. (2002)
	5.11 ± 0.05		905 ± 3	Webster (1971)
Co_2MnSi	4.96		985	Brown et al. (2000)
		5.0		Fuji et al. (1990)
		5.008		Galanakis (2005)
	4.7(375° C)			Kämmerer et al. (2004)
		5.0		Kandpal et al. (2006)
		5.0		Picozzi et al. (2002)
	4.78			Ritchie et al. (2003)
	4.95 ± 0.25			Singh et al. (2004a)
	5.1			Singh et al. (2006)
	5.0			Wang et al. (2005b)
	5.07 ± 0.05		985 ± 5	Webster (1971)
	5.07			Westerholt et al. (2005)
Co_2MnSn	4.78		829	Brown et al. (2000)
		5.03		Fuji et al. (1990)
		5.089		Galanakis (2005)
		5.0		Picozzi et al. (2002)
	5.08 ± 0.05		829 ± 4	Webster (1971)
	5.10			Zhang et al. (2005)
Co_2TiAl	0.70		148 ± 2	Souza et al. (1987)
		0.54–0.64		Ishida et al. (1982)
Co_2TiGa	0.82		128	Furutani et al. (2009)
	0.80		130 ± 2	Souza et al. (1987)
Co_2TiGe	1.94	2.0	380(5)	von Barth et al. (2010)
	1.78		386(4)	Carbonari et al. (1993)
Co_2TiSi	1.96	2.0	380(5)	von Barth et al. (2010)
	1.10		375(4)	Carbonari et al. (1993)
Co_2VAl	1.84		310(4)	Carbonari et al. (1993)
Co_2VSn	1.20		105	Carbonari et al. (1993)

Table 3.3. Theoretically predicted half-metallic other than Co-related Heusler alloys.

Compound	Magnetic moment (μ_B/unit cell)	Reference
Fe_2MnAl	1.98	Galanakis (2005)
Fe_2MnSi	2.935	Wu *et al.* (2005)
Ni_2MnSb	3.70	Rusz *et al.* (2006)
	3.882	Wu *et al.* (2005)

samples in thin-film form, especially if MR properties are to be utilized, why is there interest in growing bulk alloys if most of the applications require samples in thin-film form? One answer lies in the fact that several thin-film growth methods, such as the radio frequency magnetron sputtering method and pulsed laser deposition method, use bulk crystals or polycrystals (Giapintzakis *et al.*, 2002; Caminat *et al.*, 2004; Shen *et al.*, 2004) as sources. They add more incentives to grow bulk samples, even polycrystals. Therefore, we will discuss methods of growing both bulk samples and samples in thin-film forms.

3.3.1. *Bulk Heusler alloys*

There are two popular methods of growing Heusler alloys, namely:

- The arc-melting method for growing polycrystalline ingots.
- The tri-arc Czochralski method.

3.3.1.1. *Arc-melting method*

A typical arc-melting growth scheme (Wurmehl *et al.*, 2005) starts with a proper or a stoichiometric mixing of pure (>99.99%) constituent elements. The mixture was heated under argon atmosphere in a copper hearth cooled by water. Turning and rotating the hearth can achieve the homogeneity of the ingots. After that, the processes depend on the usages of samples, or they are pulverized and sieved (Raphael *et al.*, 2002).

Co_2FeSi samples have been grown by Wurmehl *et al.* (2005) using this method under an argon atmosphere at 10^{-4} mbar. To avoid oxygen contamination, these authors evaporated Ti inside the vacuum chamber before melting the constituent materials and carried out additional purification of the process gas. The polycrystalline ingots were annealed in a sealed near-vacuum quartz tube at 1300 K for 21 days. The samples were of the $L2_1$ structure. The compositions were checked by X-ray photoemission including electron spectroscopy of chemical analysis (ESCA) to insure the quality of crystals after the Ar^+ ion beam bombardment to remove native oxides at the surfaces. Additional checking of the structure of samples was carried out by X-ray diffraction using CuK_α or MoK_α spectra. They also carried out X-ray absorption fine structure (EXAFS) measurements to probe the short-range order of the structures.

3.3.1.2. *Tri-arc Czochralski method*

The tri-arc Czochralski method (Raphael *et al.*, 2002) is, in a sense, an extension of the arc-melting method. For single-crystal growth, the ingots are melted again by three directed arcs on a water-cooled rotating copper hearth. This three-arc scheme is the essence of the tri-arc Czochralski method. Another way is to use seed crystallites. The small seed crystals can be grown by the tri-arc method using some starting seed. For the growths of Co_2MnSi and Co_2MnGe, polycrystalline Fe has been used as starting seeds. The boule obtained this way consists of a number of grains. The crystals grown from these multigrained boules can have perfect lattice match to the desired semiconductors, such as GaAs or InAs crystal.

3.3.2. **Thin films**

Besides the commonly used growth method for HMs discussed in Chapter 2, there are two other specific methods to grow Heusler alloys in thin-film form, namely the radio frequency magnetron sputtering method (Schneider *et al.*, 2007) and the pulsed laser deposition (PLD) method (Shen *et al.*, 2004). In this section, we first revisit the MBE method in the context of Heusler alloys, then discuss the radio frequency and PLD methods.

Since there are three choices of methods to grow Heusler alloys in thin-film form, the first step is to determine which growth method is to be used. The intended use of the grown sample and properties to be examined are the main factor in this determination. The next considerations will be:

- The selection of substrates.
- The temperature at the growth, in particular the temperature of the substrates.

We shall select a few examples for each of the three methods to illustrate how the substrates are chosen and the effects of temperature.

3.3.2.1. *MBE method*

In general, the MBE method is well suited to grow low defect samples and hybrid structures in thin-film form. Achieving efficient spin injections from a magnetic sample to a semiconductor has motivated the synthesis of hybrid structures. In the following, we discuss growths of HH alloys and a hybrid structure.

Seed or buffer layers The issue of substrate is the first to be addressed. A proper choice of substrate is critical to the quality of the sample. If there is a lattice-constant matched substrate, the choice can be straightforward. However, under certain circumstances a large lattice-constant mismatch between the sample and selected substrate may be preferred. An example was discussed in Chapter 2 in the choice of MgO as a substrate for the growth of Heusler alloys. The large difference in lattice constants between the sample and substrate inevitably affects the quality of the sample. A seed or buffer layer can minimize the effect as demonstrated by Turban *et al.* (2002). They grew HH alloys using a (001)V seed layer of the order of $0.5 - 5.0\,\mathrm{nm}$ thick on (001) MgO at 400 K.

Hybrid structure with matching lattice constants Hashimoto *et al.* (2005) realized that the lattice constant of Co_2FeSi (5.658 Å) differs from that of GaAs (5.653 Å) by just 0.08%. This could lead to a better interface between the two crystals so that the efficiency of spin injection from magnetic material to semiconductor may be improved. To test that, they chose the MBE method to grow a single crystal of Co_2FeSi on GaAs (001) as a hybrid structure.

They first prepared 100 nm-thick GaAs templates in the III-V growth chamber. As-terminated c(4×4) reconstructed GaAs(001) surfaces were grown by cooling the sample to 420 °C under As_4 pressure, so that formation of defects at the top of the surface could be prevented. The GaAs sample was then transferred to an ultra-high-vacuum (UHV) chamber without any As atoms by a degas process at 580 °C. The chamber was kept at pressure 6.6×10^{-10} mbar. In growing Co_2FeSi, they started by growing layered binary alloy $Co_{0.66}Fe_{0.34}$ having a body-centered cubic (bcc) structure. By comparing the lattice constant of the alloy and taking into account the tetragonal distortion of layers, the composition was determined. They then added Si to obtain Co_2FeSi with fluxes of Fe and Co kept at the optimized values. By varying the growth temperature between 100 °C and 400 °C, they searched the optimal growth condition. This range of temperature is considered low. Therefore, these authors adopted a low growth rate at 0.1 nm/minute in order not to degrade the quality of the crystal. These conditions are essential to the growth of high-quality crystal and interface. The temperature of the Si cell (T_{Si}) varies between 1280 °C and 1335 °C for the best stoichiometry. Characterization processes and magnetization measurements confirm the quality of the single crystal and interface. Such hybrid structures have been discussed in some detail by Hirohata *et al.* (2006).

3.3.2.2. *Radio frequency magnetron sputtering method*

This method was applied in 1997 to grow thin-film forms of NiMnSb by Caballero *et al.* More recently, it has been applied to grow Co_2FeSi by Schneider *et al.* (2006) on MgO(100) and $Al_2O_3(11\bar{2}0)$ substrates, Co_2MnSi on GaAs(001) (Kohn *et al.*, 2007) and to grow Co_2FeSi on SiO_2 and MgO(001) substrates (Inomata *et al.*, 2006).

Typical setup   A typical setup is shown in Fig. 3.2. The chamber is denoted by the rectangular box and is under Ar pressure. The valve at the left corner controls the flow of the Ar gas. The pressure gauge is installed at the right corner. The base pressure in the deposition chamber is maintained under 10^{-8} mbar. The pressure is maintained by the vacuum pump, located at the bottom center of the chamber.

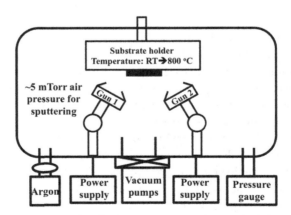

Fig. 3.2.   Schematic setup of radio frequency sputtering method (courtesy of Randy Dumas).

The growth region is composed of a substrate holder attached to the chamber. The substrate is cleaned by ultrasound with chemicals such as alcohol. It is usually heated to 500 °C or 800 °C then cooled to the growth temperature.

Two guns are placed near the middle. The targets, which are inside the guns, are cut from polycrystalline ingots obtained typically by the methods described in Chapter 2. They are controlled by the radio frequency power supply to sputter atoms onto the substrate. The typical radio frequency power is 15–100 W.

Substrate   The choice of a substrate is largely empirical. The successful ones for NiMnSb are Corning Glass (Caballero *et al.*, 1997) and for Co_2FeSi are MgO(100) and $Al_2O_3(11\bar{2}0)$ surfaces (Schneider *et al.*, 2006). The matching of lattice constants between the sample and substrate is not a major concern. For example, the lattice-constant mismatch between Co_2FeSi and MgO is 5.6%.

Temperature   The temperatures of substrates in the growth of NiMnSb are between 200 and 500 °C. The two substrates for the growth of the FH alloy are maintained at 700 °C to obtain the best quality films with thickness between 60 and 80 nm. In this case, any film grown under lower temperature conditions exhibits poorer quality — i.e., the appearance of disorder.

Schneider *et al.* (2006) obtained the best results of thin-film form of Co_2FeSi by using stoichiometric composition targets in UHV with a base pressure below 2×10^{-7} mbar and kept initially the base pressure of the chamber at 10^{-8} mbar. Then the actual growth condition is under 10^{-2} mbar of Ar pressure and at 700 °C for the substrate with the deposition rate of 5Å/s. Finally, they covered the top of the grown samples with 4 nm of Al at 350 °C to prevent oxidation. The structure of the films was determined with X-ray diffraction of the Cu-K_α line and was analyzed with an X-ray four-circle diffractometer. For the case of MgO as the substrate, they observed the (200) and (400) reflections without any impurity phases. The rocking curve has a width of 0.3°, indicating a good out-of-plane growth. When Al_2O_3 was used as a substrate, the rocking curve shows only 0.1° width.

3.3.2.3. *Pulsed laser deposition (PLD)*

This method is well suited for growing quality metallic thin-layered and multilayered structures (Shen *et al.*, 2004). It has recently been used by

Giapintzakis *et al.* (2002) to grow the HH alloy, NiMnSb, and by Wang *et al.* (2005b) to grow the FH alloy, Co_2MnSi. The targets for growing NiMnSb were pressed and sintered polycrystalline pellets. Another possibility is to obtain the pellets by radio frequency melting in an Ar gas environment. For Co_2MnSi, stoichiometric polycrystalline pellets were used.

The KrF excimer laser with the wavelength at 248 nm and pulse width of 34 ns is used as the laser source for both growths. The laser beams are incident on the rotating target at 45°. The energy of pulse is approximately 300 mJ and can have a maximum of 600 mJ. The repetition rate is 10 Hz.

Schematic diagram of setup The setup of the PLD method is shown schematically in Fig. 3.3. The chamber is shown as a large circle. The pressure in the chamber is maintained at around 10^{-9} mbar. The laser beam is provided by an excimer laser and is focused on the target by a lens. The target is composed of a bulk form of the sample. The evaporated atoms are deposited on the substrate. The sample holder is indicated by the dashed line. The thin film of the desired compound is shown as a stripe. Wang *et al.* (2005b) alternatively used a multi-chamber system combined with MBE under UHV ($\sim 10^{-13}$ mbar) conditions.

Substrate The substrate for the growth of NiMnSb using PLD is polycrystalline InAs or Si(111) surface. For the growth of Co_2MnSi films, GaAs(001) is used as an effective substrate. The substrates are cleaned. One process for cleaning the GaAs(001) substrate is as follows: the substrate is initially obtained from a commercial source. The first step is then to degas in UHV at temperature up to 580 °C. After that, the substrate is sputtered by a 0.6 keV Ar^+ ion beam with current density $4\,\mu\text{Å}/\text{cm}^2$

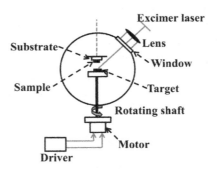

Fig. 3.3. The schematic diagram of PLD setup.

at 600 °C for 30 min, at an incident angle of 45°. The quality of the GaAs surface is checked by AES and LEED to ensure absence of surface impurities and desired reconstruction.

Temperature The temperature of the substrate for growth of NiMnSb on a semiconductor is an issue. In the MBE and radio frequency magnetron sputtering growth, better samples could be obtained with the temperature of the substrate between 350 and 400 °C. However, these materials are not suitable for making multilayer devices. The reason is that interdiffusion of atoms causes granular formations in the sample. A better growth temperature is approximately 200 °C. Both Giapintzakis *et al.* (2002) and Wang *et al.* (2005b) chose this temperature to successfully grow NiMnSb and Co_2MnSi. In addition, with the choice of this lower growth temperature they avoided the post-annealing process.

3.4. Characterization of Heusler Alloys

3.4.1. *Bulk Heusler alloys*

The crystal structure of bulk samples is determined by the Laue method. The practical rotation method is used for powder samples. Since these methods are described in elementary solid state physics text books, such as *Introduction to Solid State Physics* by Kittel (2004), we shall not provide further discussions.

At low temperature, the structure of an FH alloy is called $L2_1$. The associated space group is $Fm\bar{3}m$. The primitive cell of the $L2_1$ structure consists of four inter-penetrating fcc cubes. An $L2_1$ crystal structure is shown in Fig. 3.1.

Another related structure is the B_2 structure. In this, $X(2)$ atoms are the same as $X(1)$ atoms. Table 3.5 lists lattice constants determined either by experiments or calculations of some half-metallic Heusler alloys.

Table 3.4. The origins of the interpenetrating fcc cubes of the $L2_1$ structure. a is the outermost cube edge.

Element	Origin (a)
X(1)	(0.0, 0.0, 0.0)
X(2)	(1/2, 1/2, 1/2)
Y	(1/4, 1/4, 1/4)
Z	(3/4, 3/4, 3/4)

Table 3.5.   Lattice constants of three HH alloys and Co-based FH alloys having half-metallic properties.

Alloy	Lattice constant (Å)	Reference
NiMnSb	5.904	Van Roy *et al.* (2000)
PtMnSb	6.210	Matsubara *et al.* (1999)
PdMnSb	6.260	Matsubara *et al.* (1999)
Co_2CrAl	5.735	Yoshimura *et al.* (1985)
Co_2FeAl	5.730	Buschow and van Engen (1981)
Co_2FeSi	5.640	Wurmehl *et al.* (2005)
Co_2MnAl	5.756	Webster (1971)
	5.749	Buschow and van Engen (1981)
Co_2MnGa	5.770	Webster (1971)
Co_2MnGe	5.743	Webster (1971)
	5.75	Cheng *et al.* (2001)
Co_2MnSi	5.654	Webster (1971)
	5.66	Cheng *et al.* (2001)
Co_2MnSn	6.000	Webster (1971)
Co_2TiAl	5.85	Ziebeck and Webster (1974)
Co_2TiGa	5.85	Ziebeck and Webster (1974)
Co_2TiSi	5.849	von Barth *et al.* (2010)
Co_2TiSn	6.07	Ziebeck and Webster (1974)
Co_2VAl	5.772	Buschow *et al.* (1983)
Co_2VGa	5.779	Buschow *et al.* (1983)

3.4.2. *Thin films*

We use Co_2MnSi as an example. The AES and LEED methods are the most widely used to characterize the samples.

3.4.2.1. *Auger electron spectroscopy (AES)*

As discussed in Chapter 2, AES can probe the degree of intermixing of atoms in the substrate and film sample. One examines the evolutions of the 1070 eV Ga line, 1228 eV As line, and 780 eV Co line. The intermixing can be investigated by fitting the intensities of Ga and As lines to the following relation (Wang *et al.*, 2005b):

$$I^{Ga,As} = I_0^{Ga,As} \exp\left(-\frac{t}{\cos\phi \cdot \lambda^{Ga,As}}\right), \tag{3.1}$$

where $I^{Ga,As}$ is the intensity of either Ga or As in the presence of Co_2MnSi, and $I_o^{Ga,As}$ is the intensity of either element without Co_2MnSi. t is the thickness of Co_2MnSi. ϕ is the mean opening angle of the spherical mirror analyzer. λ^{Ga} is 15 Å for the 1070 eV Ga Auger electrons and λ^{As} is 17 Å for the 1228 eV As electrons. Data show no intermixing for Ga atoms, whereas there is a significant segregation of As to the top of Co_2MnSi. The segregations were gradually buried into the films when the thickness of the film increases.

3.4.2.2. *Low-energy electron diffraction (LEED)*

The LEED patterns of thin-film samples were compared to those of GaAs. The results confirm good quality of the epitaxial growth. However, it is important to note that the patterns are impossible to distinguish the ordered (1×1) structure from the disordered ones where same atomic species occupy several sites.

In Chapter 2, we discussed characterizations of layered structures using RHEED and STM. They have been applied to the Heusler alloys as well.

3.5. Physical Properties of Bulk Heusler Alloys

The Heusler alloys are magnetic materials and many of them were predicted to show half-metallic properties — they are the consequence of d–d or d–p interactions. One property that is of central importance is the spin polarization, P, at E_F in the metallic channel. Therefore we will focus on magnetic moments of Heusler alloys, insulating gaps, and spin polarizations of half-metallic Heusler alloys. Disorder in a sample should also be considered because of the complication of atomic arrangements in a unit cell. We divide the discussions of physical properties of Heusler alloys into three main subjects:

- Systematics of magnetic moments in half-metallic Heusler alloys.
- Individual Heusler alloys.
- Spin polarization at E_F and other magnetic properties.

3.5.1. *Magnetic moments and the Slater–Pauling rule*

The magnetic moment is of central importance in spintronics. Tables 3.1 and 3.2 list the magnetic moments of HM Heusler alloys. The question arises: is there a systematic way to estimate these magnetic moments? Most Heusler alloys contain 3d TM elements. Slater and Pauling developed an empirical rule correlating the magnetic moment per atom m of 3d elements and their alloys to the average number of valence electrons per atom, now known as the Slater–Pauling rule. We shall discuss how a Slater–Pauling type rule applies to magnetic moments in Heusler alloys.

Let the average number of the valence electrons per atom be n_v. The rule divides m (magnetic moment per atom) into two regions with Fe ($n_v = 8$) on the border. In addition, since it is often favorable for the majority d-states to be fully occupied ($n_{d\uparrow} = 5$), consistent with Hund's rule, one can define the magnetic valence $n_M = 2n_{d\uparrow} - n_v$. Malozemoff et al. (1984) and Kübler (1984) used n_M rather than n_v in their analyses of ferromagnetism in metallic systems. A plot of m vs. n_M is called the "generalized Slater–Pauling curve". Figure 3.4 shows a plot of m vs. n_M (n_v) for a series of Heusler alloys and associated TM elements obtained by Wurmehl et al. (2005). For $n_v \leq 8$, m increases with n_v, consistent with m being contributed by local moments. For $n_v > 8$, m decreases with n_v, consistent with m being contributed by mobile carriers. This is the region of itinerant magnetism.

If one defines $n_\uparrow$ ($n_\downarrow$) as the number of majority-spin (minority-spin) valence electrons per atom, then the magnetic moment per atom (in μ_B) is

$$m = n_\uparrow - n_\downarrow = (n_v - n_\downarrow) - n_\downarrow = n_v - 2n_\downarrow. \tag{3.2}$$

Fe and its binary alloys typically have approximately three valence electrons per atom in the minority channel ($n_\downarrow \approx 3$). Due to the gap in the minority channel, the half-metallic Heusler alloys have exactly

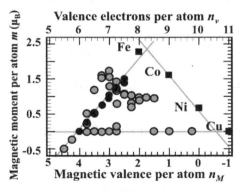

Fig. 3.4.   The Slater–Pauling curve for 3d TM elements and associated Heusler alloys. Filled circles are for Co_2-based alloys and filled gray circles are for other Heusler compounds (Wurmehl et al., 2005). n_v (n_M) is the average number of valence (magnetic) electrons per atom.

three electrons per atom in the minority channel ($n_\downarrow = 3$). Hence, from $m = n_v - 2n_\downarrow$, we have

$$m_{\mathrm{HM}} = n_v - 6. \tag{3.3}$$

For ordered HH alloys, there are three atoms/formula unit and the moment/formula unit predicted by the above rule becomes

$$M_{\mathrm{HH}} = N_v - 18, \tag{3.4}$$

where N_v is the total number of valence electrons per formula unit. Extending this argument to the FH alloys, there are 4 atoms/formula unit for a total 12 occupied bands in the $\downarrow$ spin channel, so that

$$M_{\mathrm{FH}} = N_v - 24. \tag{3.5}$$

For example, half-metallic Co_2FeSi has $N_v = 30$. The above rule then gives $M_{\mathrm{FH}} = 6\mu_B$, consistent with theory (Table 3.2).

One should bear in mind that the above rules (Eqs. (3.4) and (3.5)) determine only the magnetic moment/formula unit for a *known* HM sample. They do not predict whether a given sample is HM or not. Rather, the values of $n_\downarrow$ are determined from a band structure for each alloy as suggested by Kübler (1984).

3.5.2. *Insulating gap in half-metallic Heusler alloys*

In Heusler alloys, all the evidence shows that the majority-spin ($\uparrow$) states exhibit metallic properties while the minority-spin ($\downarrow$) channel shows insulating behavior. The insulating gap is one of the essential characteristics of having half-metallic properties. As we discussed in Chapter 1, the gap of the insulating minority-spin channel in HH alloys is generally formed between the bonding p-t_{2g} states and antibonding e_g^* states. However, as we shall see, this picture is not completely consistent with conclusions from some band structure calculations which can depend on the electronic structure methods and atomic arrangements in the unit cell.

3.5.2.1. *Half-Heusler alloys*

In the mid-1980s, Heusler alloys had already attracted much attention. van der Heidet *et al.* (1985) analyzed the ellipsometric near-infrared data and determined the insulating gaps of NiMnSb and PtMnSb as 0.7 and 0.9 eV, respectively. Later, infrared and optical absorption measurements were carried out by Kirillova *et al.* (1995) to determine the insulating gaps of NiMnSb, PdMnSb, and PtMnSb. They only reported a value of 0.4 eV for NiMnSb. The difficulty in determining such a gap is that there is a Drude contribution in metallic majority-spin states.

There are many calculations of NiMnSb and other HH alloys. Calculations of bulk properties have been carried out by a number of groups using different theoretical approaches. Aside from de Groot *et al.* (1983), who pioneered half-metallic calculations on NiMnSb, Galanakis *et al.* (2002a) used the Korringa–Kohn–Rostoker (KKR) method. The linearized MT orbital (LMTO) method was used by Kulatov and Mazin (1990), Halilov and Kulatov (1991), and Youn and Min (1995). The spin–orbit effect has been considered by Youn and Min (1995). Antonov *et al.* (1997) extended the results of de Groot *et al.* to study the magneto-optic Kerr effect in NiMnSb, PdMnSb, PtMnSb, and other FM ternary compounds. These results are listed in Table 3.3. Wang *et al.* (1994) carried out calculations on NiMnSb and PtMnSb with the atomic-sphere-approximation tight-binding LMTO method.

Most of these calculations of HH alloys show a general feature: the gap is indirect between the occupied states at the Γ point and unoccupied state at the X point. The valence states are predominantly the hybridizing p-states of the Sb and d-states of the Ni (Pd, Pt) atom. The conduction states at X is derived from antibonding d-states of the Mn atom. Therefore, p-d hybridization plays a crucial role in determining half-metallic properties of HH alloys. The spin–orbit interaction does not diminish half-metallic properties in these alloys except for PtMnSb.

Galanakis *et al.* (2002a) investigated electronic properties of a series of HH alloys using the KKR method. They did not discuss band structures for both spin states for NiMnSb because they found them to be similar to those presented by de Groot *et al.* However, they noted significantly qualitative differences between the two sets of calculations, in particular, the roles played by the Ni and Mn atoms. The most critical issue in the calculations of de Groot *et al.* is that there is no Ni-Mn d–d interaction even though Ni and Mn are nearest neighbors (nn). Furthermore, because Mn and Sb are second-nearest neighbors, de Groot *et al.* concluded that it is unlikely that d- and p-states associated with these two atoms should hybridize. In contrast, calculations performed by Galanakis *et al.* concluded that Mn d-states do mix with Ni d-states, forming states near E_F. The hybridization is purely from Sb p-states and Ni d-states around 3.5 eV below E_F because the two atoms are nn. Thus they concluded that Sb not only stabilizes the $C1_b$ structure, it also contributes states controlling the half-metallic

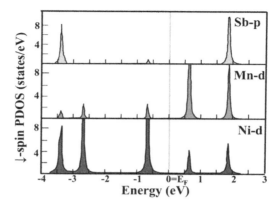

Fig. 3.5.   PDOS of ↓-spin states at Γ-point of NiMnSb (Galanakis *et al.*, 2002a).

properties. As shown in the partial DOS (PDOS) (Fig. 3.5) of the ↓-spin states, the bonding states at the Γ point are contributed from both Sb p- and Ni d-states.

With regard to the Slater–Pauling rule, Galanakis *et al.* (2002a) discussed the 18-electron rule for a binary sample to be a semiconductor: 8 electrons from the Sb and 10 electrons from the TM atom. For an HM, it should be replaced by a 9-electron rule for the insulating channel: 4 electrons from the Sb and 5 electrons from the TM element.

3.5.2.2. *Full-Heusler alloys*

There are no experimental values for the band gap of FH alloys. A few groups have calculated electronic band structures of several FH alloys using LAPW (Kandpal *et al.*, 2006) and KKR (Wurmehl *et al.*, 2005) methods.

The M value ($5.29\,\mu_B$) for Co_2FeSi calculated by the LAPW method with GGA exchange-correlation disagrees with the predicted magnetic moment/formula unit ($6.0\,\mu_B$) based on the Slater–Pauling rule, and the calculated E_F is above the gap. Only when the Hubbard U term — the on-site Coulomb repulsion — is introduced for the electron–electron correlations, do the results show HM properties. Kandpal *et al.* (2006) determined U for Co to be between 2.5 and 5.0 eV and between 2.4 and 4.8 eV for Fe. Wurmehl *et al.* (2005) found U values of 4.8 and 4.5 eV for Co and Fe, respectively. A summary of U values for different elements is listed in Table 3.6.

There is a general agreement about the characteristic of insulating gaps in FH alloys. The calculated gaps for the insulating channel are also indirect,

Table 3.6.   Summary of Hubbard U values.

Element	U (eV)	Reference
Co	2.5–5.0 (GGA)	Kandpal *et al.* (2006)
	4.8 (GGA)	Wurmehl *et al.* (2005)
	1.92 (LDA)	Kandpal *et al.* (2006)
Fe	2.4–4.8 (GGA)	Kandpal *et al.* (2006)
	4.5 (GGA)	Wurmehl *et al.* (2005)
	1.80 (LDA)	Kandpal *et al.* (2006)

Table 3.7.   The insulating gaps for Heusler alloys.

Alloy	Gap (eV)		Reference
	Exp.	Theory	
NiMnSb	0.4		Kirillova *et al.* (1995)
	0.7		van der Heidet *et al.* (1985)
		0.42	de Groot *et al.* (1983)
		0.6	Galanakis *et al.* (2000)
		0.61	Youn and Min (1995)
		0.6	Antonov *et al.* (1997)
		0.4	Halilov and Kulatov (1991)
		0.44	Kulatov and Mazin (1990)
		0.43	Wang *et al.* (1994)
PdMnSb		0.4	Antonov *et al.* (1997)
PtMnSb	0.65		Kirillova *et al.* (1995)
	0.9		van der Heidet *et al.* (1985)
		0.8	Galanakis *et al.* (2002a)
		0.91	Youn and Min (1995)
		0.2	Antonov *et al.* (1997)
		0.3	Halilov and Kulatov (1991)
		0.3	Wang *et al.* (1994)
Co$_2$MnSi		0.5	Fuji *et al.* (1990)
		1.8	Kandpal *et al.* (2006)
		1.0	Kandpal *et al.* (2006)
Co$_2$FeSi		0.8	Wurmehl *et al.* (2005)
		1.5	Kandpal *et al.* (2006)
		1.0	Kandpal *et al.* (2006)
		1.0	Kulatov and Mazin (1990)

as in NiMnSb, with the top of valence bands at the Γ point and bottom of conduction bands at the X point. The conduction states are sensitive to the U values and are derived from the antibonding states of one of the TM elements.

In Table 3.7, we summarize reported values of the insulating gap in a few Heusler alloys. For FH alloys, there are only theoretical values.

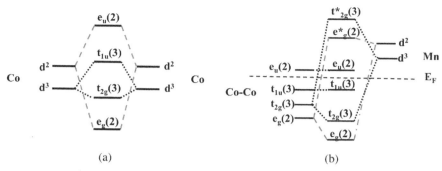

Fig. 3.6. Nature of gap states in the minority-spin channel: (a) the resulting states due to Co–Co interaction, and (b) states around E_F after Co–Co states interact with states from the Mn atom (Galanakis *et al.*, 2002b).

The nature of band-edge states of the minority-spin channel in Co_2MnSi was examined by Galanakis *et al.* (2002b). They noted a stronger d-d hybridization between the two Co atoms than between Co and Mn atoms. Figure 3.6(a) shows the resulting states due to Co–Co interaction. The five bonding states (lower group of five states at the left of Fig. 3.6(b)) then hybridize with the five states of the Mn atom, as shown in Fig. 3.6(b). They form bonding and antibonding states. The five antibonding states of the Co–Co interaction (top group of five states at the left of Fig. 3.6(b)) are located between the bonding and antibonding states resulting from the hybridization of the Co and Mn atoms, and form nonbonding states. These nonbonding states split into t_{1u} and e_u type states (states below and above E_F in the middle of Fig. 3.6(b)). Therefore, the gap states are primarily from d-states of the Co atoms.

3.5.3. *Polarization at E_F*

The polarization P at E_F is a crucial quantity if any Heusler alloy is going to be used for spintronic applications. Table 3.8 lists the measured and calculated values of P for bulk HH and Co-based FH alloys. A theory of spin-polarized positron-annihilation on NiMnSb was put forward by Hanssen and Mijnarends (1986).

Experiments were performed later by Hanssen *et al.* (1990) with the positron momentum in three different directions: [100], [110], and [111]. The [100] direction was analyzed at $T = 27$ K and the other two directions at 8 K. These experiments measured a P value of 100%, confirming half-metallic properties.

Table 3.8. P values for several Heusler alloys.

Alloy	P (%) Exp.	Theory	Temperature (K)	Reference
CoMnSb		99.0		Galanakis *et al.* (2000)
FeMnSb		99.3		Galanakis *et al.* (2000)
NiMnSb	44.0		4.2	Clowes *et al.* (2004)
	45.0		4.2	Ritchie *et al.* (2003)
		99.3		Galanakis *et al.* (2000)
PdMnSb		40.0		Galanakis *et al.* (2000)
PtMnSb		66.5		Galanakis *et al.* (2000)
Co_2MnSi	56.0		4.2	Ritchie *et al.* (2003)
	55.0			Singh *et al.* (2006)
		100.0		Fuji *et al.* (1990)
		100.0		Kandpal *et al.* (2006)
Co_2FeSi	57.0		4.2	Karthik *et al.* (2007)
$Co_2Cr_{0.02}Fe_{0.98}Si$	64.0		4.2	Karthik *et al.* (2007)

The point-contact Andreev reflection method was recently applied to $Co_2Cr_xFe_{1-x}Si$ (Karthik *et al.*, 2007). The measured P is 0.64 ± 0.01 at $x = 0.02$. As compared to the value (0.57 ± 0.01) at $x = 0$, the presence of Cr increases the P value. The authors suggest this increase is due primarily to the doping effect that improves the $L2_1$ ordering structure as concluded from their X-ray diffraction (XRD) and Mössbauer spectra.

Two groups have carried out first-principles calculations of these alloys. Both groups used the KKR method with MT potentials. Fuji *et al.* (1990) used the LSDA of von Barth and Hedin (1972) parametrized by Janak *et al.* (1975) for treating electron–electron correlation and calculated the electronic structure of crystalline Co_2MnX, where X is Al, Ga, Si, Ge, or Sn. Galanakis *et al.* (2000) performed calculations on Sb-based Heusler alloys. In Table 3.8, we include the temperature at which P is measured. Most measurements were carried out using Andreev reflection from the free surface of bulk samples.

In general, the spin–orbit interaction is small in these alloys. There are large discrepancies between experimental and theoretical results for the P values (Table 3.8). Many of the experimental authors attributed this to the presence of nonmagnetic atoms (Ritchie *et al.*, 2003; Clowes *et al.*, 2004) and the surface interrupting the tetrahedral environment around Mn and nonmagnetic atoms. de Wijs and de Groot (2001) calculated ideal surfaces of NiMnSb and showed that they are not HMs, due to symmetry breaking

Table 3.9. Calculated magnetic moments of TM elements for some HH alloys, XYZ. X is either Ni or Co. Y denotes V, Cr, Mn, or Fe. Z is In, Sn, Sb, or Te. "Void" is for a site not occupied by any atom. These values were obtained by Galanakis *et al.* (2002b) except CoVSb which was calculated by Tobola *et al.* (1998).

Compound	Ni (μ_B)	Mn (μ_B)	X (μ_B)	Void (μ_B)	Total (μ_B)
NiMnIn	0.192	3.602	−0.094	0.003	3.704
NiMnSn	0.047	3.361	−0.148	−0.004	3.256
NiMnSb	0.264	3.705	−0.060	0.052	3.960
NiMnTe	0.467	3.996	0.101	0.091	4.656
NiCrSb	0.059	2.971	−0.113	0.059	2.976
NiFeSb	0.404	2.985	−0.010	0.030	3.429
NiVSb	0.139	1.769	−0.040	0.073	1.941
CoVSb	−0.126	1.074	−0.021	0.038	0.965
CoCrSb	−0.324	2.335	−0.077	0.032	1.967
CoMnSb	−0.132	3.176	−0.098	0.011	2.956
CoFeSb	0.927	2.919	−0.039	0.012	3.819

at the surface. However, when the tetrahedral environment is restored, the half-metallicity is recovered.

3.5.4. *Magnetic moments*

Magnetic moments of Ni- and Co-based HH alloys were calculated by Galanakis *et al.* (2002b). Table 3.9 lists total magnetic and local moments for some pnictides and chalcogenides. These calculated results are based on experimental lattice constants for each compound. CoVSb was also studied theoretically by Tobola *et al.* (1998). They concluded that it is an HM with a spin moment of $0.965\,\mu_B$. Its experimental lattice constant is 5.801 Å. For all of the compounds studied in this paper, the interesting feature is the magnetic moment vs. valance of the low-valence TM elements. For Cr replacing Mn and V replacing Cr, the magnetic moment reduces by $1.0\,\mu_B$ — indicating the existence of local moments. For Fe replacing Mn, due to the fact that majority d-states are now filled, an extra electron fills the minority-spin states and thus half-metallic properties are lost. Galanakis *et al.* (2002b) also remarked on replacing Sb by Te or Sn, which destroys the half-metallicity. For Te, the additional electron puts E_F above the gap of the minority-spin channel while for Sn, E_F falls below the gap.

Table 3.10 lists the effect of the lattice parameter on magnetic moments of a few HH alloys calculated by Şaşıoğlu *et al.* (2005b), where $a_{I[exp]}$

Table 3.10. Comparison of computed magnetic moments of NiMnSb and CoMnSb under compression. "Void" corresponds to the missing atom in the $L2_1$ structure.

Compound	Lattice constant (Å)	Ni or Co (μ_B)	Mn (μ_B)	Sb (μ_B)	Void (μ_B)	Total (μ_B)
NiMnSb–$a_{I[\exp]}$	5.93	0.20	3.85	−0.09	0.04	4.00
NiMnSb–a_{II}	5.68	0.32	3.68	−0.05	0.05	4.00
NiMnSb–a_{III}	5.62	0.33	3.64	−0.04	0.05	3.97
CoMnSb–$a_{I[\exp]}$	5.87	−0.32	3.41	−0.11	0.02	3.00
CoMnSb–a_{II}	5.22	0.45	2.57	−0.06	0.04	3.00
CoMnSb–a_{III}	5.17	0.48	2.52	−0.05	0.04	2.99

corresponds to the experimental bulk lattice constant. For both NiMnSb and CoMnSb, E_F lies in the lower part of the insulating gap. The use of the lattice constant a_{II} shifts E_F to the upper edge of the insulating gap. Further compression of 1%, to a_{III}, places E_F slightly above the gap. The effect of the lattice parameter on the local magnetic moment is determined within the atomic sphere. For example, for NiMnSb, the lattice contraction increases the hybridization between Ni and Mn atoms and increases the magnetic moment of the Ni atom while decreasing the moment of the Mn atom. In CoMnSb, which has a larger insulating gap, the moments on the Co and Mn atoms are antiparallel. As a result, the transition of E_F to the upper edge of the insulating gap requires a large lattice contraction of ∼11%. The magnetic moment at each atom is sensitive to the lattice parameter while the total moment is relatively insensitive.

3.5.5. *Curie temperature T_C*

Curie temperature T_C of a Heusler alloy is of central interest for RT spintronic device applications. Most of the measured T_C values have been obtained by inverting the magnetic susceptibility χ. A summary of T_C values for Heusler alloys is given in Table 3.11. It is most encouraging to note that so many Heusler alloys have T_C well above RT.

3.5.6. *Other magnetic properties*

XMCD spectra of NiMnSb were studied experimentally by Kimura *et al.* (1997) and Yablonskikh *et al.* (2000, 2001) and theoretically by Galanakis *et al.* (2000) (Fig. 3.7). The difference between the two experiments is that absorption spectra (Kimura *et al.*, 1997) and emission spectra (Yablonskikh *et al.*, 2000, 2001) were measured, respectively. The measured quantity in

Table 3.11. T_C for Heusler alloys.

Alloy	T_C (K)	Reference
NiMNSb	728	Otto *et al.* (1989)
	728	Ritchie *et al.* (2003)
Co$_2$MnGa	694	Brown *et al.* (2000)
Co$_2$MnGe	905	Brown *et al.* (2000)
Co$_2$MnSi	985	Brown *et al.* (2000)
	985	Ritchie *et al.* (2003)
Co$_2$MnSn	829	Brown *et al.* (2000)
Co$_2$FeGa	>1100	Brown *et al.* (2000)
Co$_2$FeSi	1100	Wurmehl *et al.* (2005)
Co$_2$TiSn	359	Brown *et al.* (2000)
	370	Souza *et al.* (1987)

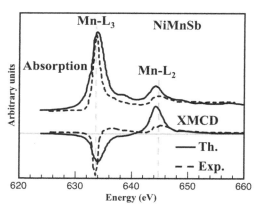

Fig. 3.7. Experimental (Kimura *et al.*, 1997) and theoretical (Galanakis *et al.*, 2000) absorption and XMCD spectra of Mn $L_{2,3}$ lines in NiMnSb.

X-ray absorption spectra is the difference of intensities with the photon helicity (spin) parallel (I_+) and antiparallel (I_-) to the magnetization. The $L_{2,3}$ excitations from Mn $2p$ core states were measured.

Physically, when a core electron in the $2p$ level is excited, the absorption process is governed by selection rules of the transition to $3d$ states of the Mn atom. If the sample is half-metallic, the final states of the absorption process are majority-spin states at E_F. The insulating minority-spin channel is not expected to have any states to contribute. As the photon energy increases to reach conduction states of the minority-spin channel, it is then possible to obtain a finite intensity I. The difference $I_+ - I_-$ can therefore reveal half-metallic properties. The XMCD Mn-$L_{2,3}$ edge spectra of NiMnSb are shown in Fig. 3.7. The negative part centred around 634.0 eV (L_3 line) is stronger than the positive contribution centred at 645.0 eV (L_2 line).

The XMCD absorption spectrum of Mn-based HH alloys can be calculated similarly to any optical transitions except with the additional contribution from the spin polarization.

$$\varepsilon_{2,\pm}(\omega) = \frac{2\pi}{\hbar} \sum_{m_{j\pm}} \sum_{n,\mathbf{k}} \langle j_\pm m_{j\pm} | \mathbf{e}_\pm \cdot \mathbf{p} | n\mathbf{k} \rangle \langle n\mathbf{k} | \mathbf{p} \cdot \mathbf{e}_\pm | j_\pm m_{j\pm} \rangle$$

$$\times \, \delta(\hbar\omega - E_{n\mathbf{k}} + E_{j_\pm}), \tag{3.6}$$

where $|j_\pm m_{j\pm}\rangle$ is an initial core state having energy $E_{j_\pm}$, $|n\mathbf{k}\rangle$ is a final state with energy $E_{n\mathbf{k}}$, $\mathbf{p}$ is the electron momentum, and $\mathbf{e}_\pm$ are polarizations of photons. The δ-function enforces conservation of energy. The two peaks at 634 eV and 644.5 eV calculated by Galanakis *et al.* (2000) agree well with experimental results. The linewidth of the lower energy peak does not agree due primarily to the fact that theoretical results were from ideal crystals.

The emission process involves an electron near E_F dropping into a core-hole state. One measures the emitted X-ray. This process is more complicated than absorption because the presence of an equilibrium hole in the core region can affect energies of valence states. The corresponding structures in the measured spectra are at 640.5 eV and 652 eV, respectively (Yablonskikh *et al.*, 2000, 2001).

Another element-specific, surface-sensitive magnetic technique — the spin-resolved appearance potential spectroscopy (SRAPS) — has been applied to probe the surface magnetization of the NiMnSb(001) surface (Kolev *et al.*, 2005). The basic idea is related to core–core–valence Auger transitions (Hörmandinger *et al.*, 1988). In Fig. 3.8, the basic processes are illustrated. The direct process has core–core transition between states 3 and 1 and valence–valence transition between states 4 and 2. The exchange process is characterized by valence–core transition from state 4 to state 1 and core–valence transition from state 3 to state 2.

Kolev *et al.* (2005) formulated a simple expression for the intensity of SRAPS:

$$I_\sigma(E) \sim \int_0^{E+E_c} \sum_{\kappa\mu,\gamma\delta} n_{\kappa\mu}(E')n_{\gamma\delta}(E+E_c-E')W^\sigma_{\kappa\mu,\gamma\delta}(E', E+E_c-E')dE',$$

$$\tag{3.7}$$

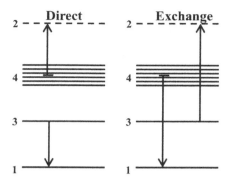

Fig. 3.8.   Schematic energy level diagram for core–valence Auger transitions. Numbers 1 and 3 are core states; 2 and 4 are valence states (Hörmandinger *et al.*, 1988).

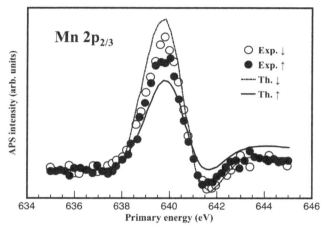

Fig. 3.9.   Comparison of theoretical (bulk) and experimental (surface) APS spectra for the ↓- and ↑-spin states (Kolev *et al.*, 2005).

where the first Greek letter labels the relativistic spin–orbit quantum number of the state and the second Greek subscript denotes the magnetic quantum number. n is the LDOS at the atom of interest and can be obtained from first-principles calculations. W is the effective spin-dependent cross section. The measured results show that the SRAPS amplitude is dependent on the location at the surface for the X-ray to create holes. The measured surface SRAPS spectra are compared to bulk SRAPS spectra in Fig. 3.9. The key feature in Fig. 3.9 is the ratio of amplitudes for the majority- and minority-spin channels. The experimental result is approximately three times smaller than the theory based on the bulk model. This indicates the reduction of spin polarization at the surface. This result agrees with those of other experiments.

The XMCD emission spectra of $L_{2,3}$ in Co_2MnSb were measured by Yablonskikh *et al.* (2000, 2001). The two structures relating to the majority- and minority-spin channels are at 638 eV and 648 eV, respectively. The energies of these structures depend on the photon energy to excite a hole in the core region. The higher energy peak does not appear until the exciting photon has an energy of 652 eV.

3.5.7. *Disorder in Heusler alloys*

Because of complex crystal structures for HH and FH alloys, disorder in the samples is inevitable. We first comment briefly on probing the disorder in these alloys and discuss the nature of disorder in HH and FH alloys.

3.5.7.1. *Experimental probes*

There are two methods to probe local disorder in Heusler alloys: Mössbauer spectroscopy including the determination of the hyperfine field (Khoi *et al.*, 1978), and spin echo nuclear magnetic resonance. Neutron and X-ray scattering are methods for determining the degree of disorder.

Mössbauer spectroscopy   This method makes use of the recoil speed of isotopes while the nuclei emit or absorb γ-ray in order to probe the presence of disorder from linewidths of the radiation. Alloys composed of Fe, Mn, and Sn can be readily probed. By including the interaction between the nuclear spin I_z and effective magnetic field H_e at the site, the hyperfine field $H_{hf} = -\gamma I_z H_e$ can be determined, where γ is the gyromagnetic ratio. For example, ^{119}Sn Mössbauer spectroscopic measurements on Co_2MnSn with Fe or Cr substituting Mn at RT have been carried out by Zhang *et al.* (2005). Mössbauer spectra for $Co_2Mn_{1-x}Fe_xSn$ is shown in Fig. 3.10. There is a well-defined peak centered at about 2 mm/s for $x = 0$. When $x \geq 0.2$, there is a doublet structure that appears on the peak, suggesting a small percentage of Sn forming a non-Heusler phase. Wurmehl *et al.* (2006a) used ^{57}Co to probe the disorder in a powdered sample of Co_2FeSi at 85 K. They obtained a sextet pattern with an isomer shift of 0.23 mm/s. This pattern is typical for magnetically ordered systems. They also observed an ^{57}Fe line with a width of 0.15 mm/s, indicating a well-ordered sample. This width is comparable to the 0.136 mm/s width for α-Fe at 4.2 K. More recently,

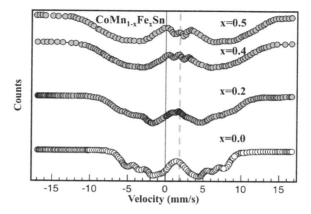

Fig. 3.10. Mössbauer spectra of $Co_2Mn_{1-x}Fe_xSn$ (Zhang *et al.*, 2005).

Karthik *et al.* (2007) reported Mössbauer spectra of $Co_2Cr_xFe_{1-x}Si$. Their results also show a sextet pattern. For $x \geq 0.2$, there is a doublet structure.

Spin echo nuclear magnetic resonance It is theoretically possible to calculate the hyperfine field at a nucleus if the chemical environment is known. Consequently, direct measurements of nuclear magnetic properties are effective to determine the disorder in the samples. The nuclear magnetic resonance (NMR) frequencies are of the order of MHz. The linewidths of NMR peaks reveal the degree of disorder in the samples. For off-stoichiometric and substitutional alloys, two quantities are measured: the hyperfine field on impurity atoms, and positions of NMR satellite lines originating from host atoms with nearest neighbor and second-nearest neighbor impurity atoms, in particular for low concentrations. The total local effective magnetic field at the i-th atom is

$$H_i = h_{oi}\mu_o + h_{\langle nn \rangle i} \sum \mu_{\langle nn \rangle} + h_{\langle sn \rangle i} \sum \mu_{\langle sn \rangle} + \cdots, \qquad (3.8)$$

where o means at the i-th site, $\langle nn \rangle$ and $\langle sn \rangle$ denote the first- and second-nearest neighbors, respectively. h_{ji} is a parameter. Let A_i be the hyperfine coupling constant of the i-th atom. We can calculate h_{ji}/A_i to compare contributions at the i-th atom due to different neighbors.

The spin echo experiments provide more information about the disorder in a sample. Wojcik *et al.* (2002) found three lines at 217, 263, and 199 MHz in NiMnSb. These lines were identified from ^{123}Sb, ^{55}Mn, and

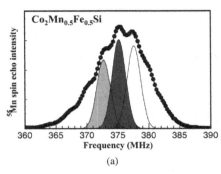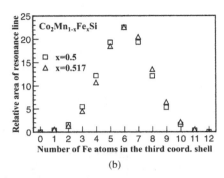

(a)　　　　　　　　　　　　　　　　　(b)

Fig. 3.11. (a) [55]Mn spin echo intensity as a function of frequency in $Co_2Mn_{0.5}Fe_{0.5}Si$. The distribution of Fe atoms in the third coordination shell of the [55]Mn is also given; (b) relative area of resonance line vs. number of Fe atoms in the third coordination shell of [55]Mn. Shown are the results for the $Co_2Mn_{0.5}Fe_{0.5}Si$ compound with ideal stoichiometry $x = 0.5$ (squares) and optimized stoichiometry $x = 0.517$ (triangles) (Wurmehl *et al.*, 2007).

[121]Sb, respectively. The authors also suggested that there was twinning in their samples.

Wurmehl *et al.* (2007) applied spin echo technique to $Co_2Mn_{1-x}Fe_xSi$. By defining the probability for the local environment surrounding an Fe atom to have n Mn atoms and $12 - n$ Fe atoms as

$$P(n, x) = \frac{N!}{(N - n)!n!}(1 - x)^{N-n}x^n, \tag{3.9}$$

$P(n, x)$ can be directly compared to the intensity of spin echo measurements. In this particular experiment, [55]Mn spin echo intensity as a function of frequency in $Co_2Mn_{0.5}Fe_{0.5}Si$ (Fig. 3.11(a)) and relative area of the resonance line vs. number of Fe atoms in the third coordination shell of [55]Mn (Fig. 3.11(b)) were obtained.

Other methods Neutron diffraction experiments estimate less than 10% atomic disorder in NiMnSb (Helmholdt *et al.*, 1984). Kautzky *et al.* (1997) carried out XRD measurements on two single phase (111) oriented thin films of PtMnSb and found 10% disorder in each film. The possible disorders in two different films are given in Table 3.12. These results were obtained by measuring the (111), (333), and (444) peaks, then fitting their integrated intensities with structure factors characterizing different types of disorder. Note that the vacant site $(X(2))$ in $C1_b$ sublattice can easily cause disorder.

Table 3.12. Disorder determined from XRD measurements on two films of PtMnSb. Each value in parentheses is the fractional occupancy of sites on a particular $C1_b$ sublattice which are occupied by that atom type.

Film thickness (Å)	Site (see Fig. 3.1)	Fractional occupancy
1090	X(1)	Pt(0.72) Pt(0.77)
	X(2)	Pt(0.15) Pt(0.10)
		Mn(0.15) Sb(0.10)
	Y	Mn(0.79) Mn(0.94)
	Z	Sb(1.00) Sb(0.90)
543	X(1)	Pt(0.80) Pt(0.83)
	X(2)	Pt(0.11) Pt(0.07)
		Mn(0.11) Sb(0.07)
	Y	Mn(0.89) Mn(1.00)
	Z	Sb(1.00) Sb(0.92)

Table 3.13. Three defect models studied by Orgassa *et al.* (1999, 2000).

Defect structure	Type	Site occupation X(1)	X(2)	Y	Z
X(1) $\leftrightarrow$ Y	1	$Ni_{1-x}Mn_x$		Ni_xMn_{1-x}	Sb
X(1)Y $\leftrightarrow$ X(2)	2	Ni_{1-x}	Ni_xMn_x	Mn_{1-x}	Sb
YZ $\leftrightarrow$ X(2)	3	Ni	Mn_xSb_x	Mn_{1-x}	Sb_{1-x}

3.5.7.2. *Theoretical investigations of disorder*

Half-Heusler alloys Structural defects in NiMnSb were investigated theoretically by Orgassa *et al.* (1999, 2000) using the KKR method with the coherent potential approximation at the experimental lattice constant, 5.927 Å. They considered three defect models. Recall that there are four interpenetrating fcc cubes in the crystal structure. The cube related to X(2) site in NiMnSb is not occupied. In Table 3.13, three models labeled as types 1, 2, and 3 are listed. In all cases the composition is kept stoichiometric. The question is whether the disorder will destroy the half-metallic properties. These authors calculated P values for the three defect models with 10 and 15% of disorder for each structure. The results are summarized in Table 3.14 which shows that in general the value of P decreases monotonically with increasing disorder. The worst-case scenario is type-3 disorder. Physically, when the unoccupied X(2) site is occupied by a TM element, the d–d interaction is changed in such a way as to diminish the half-metallicity.

Table 3.14.   The effects of disorder on P in NbNiSb studied by Orgassa *et al.* (1999).

Defect type	Disorder (%)	P (%)
1	5	52
	10	29
2	5	67
	10	31
3	5	24
	10	10

Table 3.15.   Formation energy ΔE and total magnetic moment M in a unit cell for the four types of disorder calculated by Picozzi *et al.* (2004).

	ΔE (eV)		M (μ_B)	
Disorder	Co_2MnSi	Co_2MnGe	Co_2MnSi	Co_2MnGe
Co antisite	0.80	0.84	38.01	38.37
Mn antisite	0.33	0.33	38.00	38.00
Co-Mn swap	1.13	1.17	36.00	36.00
Mn-Si swap	1.38		40.00	

Full-Heusler alloys   Picozzi *et al.* (2004) investigated the effects of disorder in a 32-atom supercell due to the exchange of Mn and Co atoms and antisites in Co_2MnSi and Co_2MnGe. The Mn antisite is defined as a Co site which is occupied by a Mn atom, while the Co antisite is a Mn site occupied by a Co atom. The method of calculations was the LAPW method with GGA exchange-correlation. The MT radius for all atoms was 1.11 Å. The formation energies were computed as follows:

$$\Delta E = E_{\text{def}} - E_{\text{ideal}} + n_{Mn}\mu_{0,Mn} + n_{Co}\mu_{0,Co} + n_X\mu_{0,X}, \qquad (3.10)$$

where the first term is the total energy with disorder, the second term is the total energy without disorder, n_i is the number of atoms transferred to and from a chemical reservoir of the i-th element, $\mu_{0,i}$ is the corresponding chemical potential, and X denotes Si or Ge. The chemical potentials of Mn, Co, and X were determined from fcc antiferromagnetic Mn, hcp FM Co, and diamond-structure Si and Ge. The formation energies and the total magnetic moments in the supercell are summarized in Table 3.15. In the ideal (perfectly ordered) case, the total magnetic moment is $40\,\mu_B$. The Mn antisite has the smallest formation energy. Picozzi *et al.* suggested that it can be easily formed during growth, especially in the tri-arc Czochralski method for Co_2MnSi at $1523 K$. Experiments performed by Raphael *et al.* (2002) show there are $0.36 \times 10^{22} cm^{-3}$ Mn antisites.

3.6. Physical Properties of Heusler Alloys in Thin-film Form

In this section, we shall discuss the physical properties of a few Heusler alloys in thin-film forms: first, we shall discuss the most stable HH alloy, NiMnSb, then some FH alloys, including the most-studied Co_2MnSi.

3.6.1. *NiMnSb*

3.6.1.1. *Stability of structure and half-metallicity*

Based on the theoretical studies of bulk properties of HH alloys, the NiMnSb-type structure is the most stable among all HH alloys, with respect to an interchange of atoms (Larson *et al.*, 2000). Orgassa *et al.* (1999, 2000) showed that the half-metallicity in NiMnSb is not destroyed by a few percent disorder. Galanakis *et al.* (2002b) attributed the gap in the non-metallic channel to the d–d interaction between Ni and Mn atoms, which differs from the explanation in terms of the d–p interaction between Mn and Sb atoms proposed by de Groot *et al.* (1983). However, Galanakis *et al.* argued that the presence of Sb atoms is crucial to stabilize the structure because NiMn does not favor any open structure.

3.6.1.2. *Spin polarization*

The NiMnSb films were grown on GaAs(001) by Van Roy *et al.* (2000) using the MBE method and on MgO(001) and Si(001) by Schlomka *et al.* (2000), using argon-ion sputtering onto water-cooled targets. The thickness of the films in the first experiment was 260 to $350 nm$. The single-crystal samples were grown in the [001] orientation. In the second experiment, the thickness ranged from 1.0 to $80 nm$. The surfaces showed roughness. As the temperature was increased from 150 to $250 °C$, the roughness was increased by a factor of 5. These experiments found that the NiMnSb films were not half-metallic.

On the other hand, Zhu *et al.* (2001) grew polycrystalline samples on Si substrates using the e-beam evaporation method of Kabani *et al.* (1990) and carried out spin-resolved photoemission measurements. They obtained 40% spin polarization which is smaller than an earlier result of 50% obtained by Bona *et al.* (1985).

To explain the much smaller measured polarization as compared to the theoretically predicted value, two suggestions were made. The first is that the gap in the minority-spin channel is less than the calculated value of 0.5 eV (Kang *et al.*, 1995). The second suggestion is that surface segregation occurs (Park *et al.*, 1998a).

Ristoiu *et al.* (2000b) and Komesu *et al.* (2000) examined the effects of surface composition on the polarization. They grew MgO(110)/ Mo(100)/NiMnSb(100) thin films by sputtering, and capped the samples with a 1000 Å Sb layer. Inverse photoemission experiments were carried out to measure the polarization with capping layers removed from the samples. The P value was found to be $67 \pm 9\%$ (Ristoiu *et al.*, 2000b). The polarization decreased as the thickness of the Sb layer increased (Komesu *et al.*, 2000).

Beside the issue of determining experimentally the polarization P at Fermi level, there is a transition in NiMnSb from an HM to a normal FM at about 80 K (Hordequin *et al.*, 2000; Borca *et al.*, 2001); the temperature is referred to as T^*. One possible explanation for the transition is that E_F is located very close to the bottom of the conduction bands in the minority-spin channel (Fig. 3.12). As the temperature increases, electrons at E_F in the majority-spin channel can be thermally excited to the conduction band edge with their spin orientations flipped from the ↑ state to the ↓ spin state. As the conduction band of the ↓ spin channel is occupied, the half-metallicity is lost. The temperature for this to happen in NiMnSb is 88 K (i.e., $\delta \leq 88$ K). Consequently, it is unlikely that NiMnSb can be used for fabricating the RT spintronic devices.

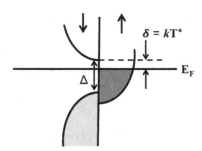

Fig. 3.12. A schematic diagram of the DOS of an HM with the majority-spin channel showing metallic behavior. The Fermi energy of the Heusler alloy is indicated as E_F and is located right below the conduction band edge by a spin-flip gap δ. Δ is the fundamental gap of the insulating channel.

3.6.1.3. *Surface and interface effects*

Using angle-resolved X-ray photoemission spectroscopy (ARXPS), Ristoiu *et al.* (2000b) suggested that it is possible to have MnSb- or Mn-rich surface layers on top of NiMnSb thin-film samples depending on how the surfaces were prepared. Ristoiu *et al.* (2000a) found that Mn segregations (Mn-rich) and vacancies in Ni layers, instead of MnSb, terminate the surface of the films by preparing the samples with sufficient annealing for stoichiometric surfaces. The effect of Mn-rich surface structure is the cause of the reduction of spin polarization.

A theoretical study by Jenkins (2004) was performed to examine the effects of the hexagonal phase of MnSb formed at the surface of NiMnSb. They used the plane wave pseudopotential method to examine MnSb-terminated (001) (1 × 1) NiMnSb surface. The supercell consists of a slab with five MnSb/Ni bilayers. A slight buckling of 0.06 Å in the top MnSb layer was found by relaxing the surface — the Sb atoms relax outward relative to their bulk positions, while the Mn atoms relax inward. The second layer buckles by 0.09 Å in the reverse sense to the top layer. The third layer has 0.03 Å buckling. The calculated magnetic moment of the Mn atom at the surface is $4.21\,\mu_B$, increased by 9% as compared to the bulk value. The symmetry breaking due to the presence of surface introduces surface states, making bands of the minority-spin channel cross E_F. Therefore, in the surface region, the sample is metallic. The occupied surface states at the Γ point of the surface BZ originate from Mn $d_{x^2-y^2}$ orbitals. The surface states above E_F are derived from d_{xz} orbitals of the Mn atoms mixed with p_x states of the Sb atoms. The surface band structures along Γ-K and Γ-M with different terminations at the surface are shown in Fig. 3.13. In addition, Wojcik *et al.* (2002) observed MnSb in epitaxially grown NiMnSb on a GaAs(001) surface. For Sb-poor samples, the MnSb inclusions disappear.

de Wijs and de Groot (2001) designed a NiMnSb/CdS supercell with different interface structures and used the plane wave pseudopotential method to examine the interface structure on the spin polarization. They found that (100) interface diminished the spin polarization. On the other hand, the (111) interface with interface composed of the Sb and S atoms as shown in Fig. 3.14 restores half-metallicity. It was also found that (100) and (111) surfaces of NiMnSb do not exhibit half-metallic properties.

The magnetic properties of half-metallic NiMnSb and CoMnSb surfaces have been calculated by Galanakis (2002b). The results are summarized

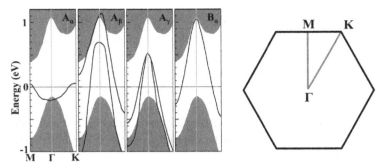

Fig. 3.13. Surface band structure of the NiMnSb(111) surface (Jenkins, 2004). The upper surface terminated by Mn is denoted as A_α, by Ni as A_β, and by Sb as A_γ. The bottom surface terminated by Sb is denoted by B_α.

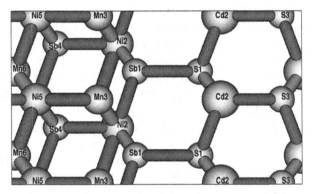

Fig. 3.14. Structure of the half-metallic NiMnSb(111)/CdS(111) interface (de Wijs and de Groot, 2001).

in Table 3.16, along with the corresponding bulk values. In general, the magnetic moment at the surface is increased mainly due to the presence of TM elements with increased local moments at the surface.

Magnetization and Kerr rotation measurements were carried out on thin films of NiMnSb by Kabani *et al.* (1990). The results of the hysteresis loop for a film sample of 220 nm measured at 4.2 K show significant anisotropy (Fig. 3.15). The origin of the anisotropy is not identified. The magnetic moment of $3.9 \pm 0.2 \, \mu_B$ per formula unit was determined. These authors also measured the Kerr rotation at RT on the same film using the MOKE method. A laser source with a wavelength of 632.8 nm was incident on the film surface at 45°. The magnetic field was swept

Table 3.16.   Spin moments in μ_B for NiMnSb and CoMnSb compounds in the case of: (i) bulk compounds; (ii) the Mn and Sb atoms in the surface; and (iii) the Ni- or Co-terminated surfaces. The "total" moment denotes the sum of moments in the surface and subsurface layers that include Sb and vacancy (Galanakis, 2002b).

	NiMnSb			CoMnSb		
Interface	Ni	Mn	Total	Co	Mn	Total
Bulk	0.26	3.70	3.96	−0.13	3.18	2.96
(001)MnSb	0.22	4.02	4.19	−0.06	3.83	3.65
(001)Ni	0.46	3.84	4.30			
(001)Co				1.19	3.31	4.43

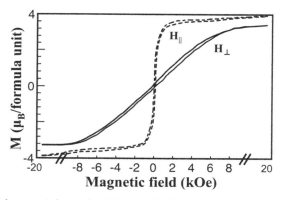

Fig. 3.15.   The hysteresis loop of a 220 nm NiMnSb film at 4.2 K (Kabani *et al.*, 1990).

between ± 3.5 kOe. At the limiting value of the magnetic field and in the polar configuration, the Kerr rotation is $0.06°$ as compared to the bulk value of $0.10°$. It seems that the film has magnetic properties close to those of the bulk.

3.6.2.   *Co₂MnSi*

Among all Heusler alloys, bulk Co_2MnSi has several appealing properties. It is predicted to be an HM with a large insulating gap of about 0.4 eV in the $\downarrow$ spin channel (Fuji *et al.*, 1990). It possesses the second highest T_C (985 K) among all Heusler alloys (Brown *et al.*, 2000). For spintronic applications, samples in thin-film form are preferred. Therefore, it has attracted much study in thin-film forms. GaAs has been used as the substrate. Amorphous Al-O and MgO have been used as the barrier in tunnel junctions. The high polarization deduced from the magnetic tunnel junction made of Co_2MnSi shows great potential for spintronic devices.

3.6.2.1. *Growth in thin-film form*

Co_2MnSi in thin-film form is among the most-studied FH alloys. It has been grown on a GaAs(001) surface by PLD (Wang *et al.*, 2005b), inductively coupled plasma-assisted magnetron sputtering and dc magnetron sputtering (MS) (Kohn *et al.*, 2007). We shall discuss in detail the growth by PLD and dc MS methods because of some unique requirements during the growth. Toward device applications, we shall discuss MTJs, in particular $Co_2MnSi/Al-O/Co_2MnSi$, which were fabricated using the inductively coupled plasma-assisted MS method by Sakuraba *et al.* (2006).

In 2005, Wang *et al.* reported the use of polycrystalline Co_2MnSi pellet targets to grow thin-film forms of Co_2MnSi by PLD on GaAs(001) substrates. The lattice constant of GaAs (5.65 Å) matches that of Co_2MnSi. These substrates were commercially available and were degassed in UHV up to 580 °C. Finally, the surfaces of substrates were sputtered for 30 minutes at 600 °C by an Ar^+ ion beam with energy at 0.6 keV. The beam was incident at 45° with current density $4.0 \, \mu A/cm^2$. A KrF excimer laser of 248 mm wavelength was used with 34 ns pulse width, 300 mJ energy, and 10 Hz repetition rate. The pressure inside the chamber was kept under 5.0×10^{-11} mbar. Previous experience showed difficulties in growing quality Heusler alloys on semiconductor surfaces at high deposition temperatures (>450 K). The present growth was carried out with substrate temperature at 450 K. Quality thin films of around 60 Å were obtained. Monitoring RHEED oscillations provided the evidence of the film quality. Regular oscillations started at the third layer and continued up to ten layers. The lack of oscillations at the first two layers was attributed to the poor layer-by-layer growth and the possibility of interface mixing. The characterization of the quality of the films by AES has been given in Section 3.3.

More recent growth details were given by Kohn *et al.* (2007). The GaAs (001) substrates were positioned below the targets on a Ta strip heater. The substrates were chemically cleaned and annealed at 595 °C for 10 minutes to remove oxides and to have a 4×2 reconstruction. The base pressure in the deposition chamber was 2.67×10^{-9} mbar. The temperature was then lowered to 380 °C and the system was pumped for one-and-a-half hours to remove the As atoms completely — avoiding the formation of Mn_2As compound. The rate of deposition was 0.10 nm/s under an Ar pressure of 32 mbar. Energy dispersive X-ray (EDX) analysis in a scanning electron microscope was carried out and determined stoichiometric samples within 1.5 at.%. X-ray diffraction patterns showed the films having single-phase (001) orientation with a lattice constant of 5.63 ± 0.01 Å. Film thicknesses of 15 to 260 nm were obtained. There was intermixing between Mn and As atoms in the films up to 15 nm. For thicker films, the $L2_1$ structure is the main polycrystalline phase.

3.6.2.2. *Magnetic properties*

X-ray absorption and magnetic circular dichroism The spin polarization, P, at E_F is the quantity of interest and is determined by the magnetic moments of TM elements. The X-ray absorption spectra (XAS) and XMCD methods, combined with sum rules, are widely used for determining magnetic moments. Typically, the $L_{2,3}$ absorption edges of the Mn and Co atoms are used. The sum rule relates the integrated intensity difference of the two lines to the orbital and spin moments of the atoms, corrected for the incomplete degree of circular polarization and angle of incidence.

The XAS spectra for Co-$L_{2,3}$ of a Co_2MnSi film of 17 Å thickness at RT obtained by Wang *et al.* (2005b) are shown in Fig. 3.16. The main features are as follows:

- For the Co atom, XAS shows lines at 776.5 eV and 791.5 eV corresponding to $2p_{3/2} \rightarrow 3d$ (first peak, L_3) and $2p_{1/2} \rightarrow 3d$ (second peak, L_2) transitions, respectively. The edges of $L_{2,3}$ lines for the Mn atom are at 638.5 eV and 651.4 eV.
- The linewidths of the Mn atom are narrower than those of the Co atom.

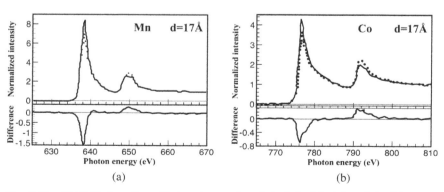

Fig. 3.16.   XAS Co-$L_{2,3}$ lines (first and second peaks) of the Mn (a) and Co (b) atoms from a Co_2MnSi film of 17 Å thick (Wang *et al.*, 2005b).

- The high-energy line of Mn exhibits a doublet structure at 649.4 eV and 650.8 eV.
- There is some variance with respect to the thickness of the films. For example, the doublet structure of the Mn L_2 line is smeared in a thicker (45 Å) film.
- In contrast to the L_3 line of the Co atom in other Heusler alloys such as Co_2TiSn, Co_2ZrSn, and Co_2NbSn (Yamasaki *et al.*, 2002), showing multiple structures, this one does not exhibit any such structures.

The narrowness of the Mn lines can be attributed to the number of Mn and Co atoms in the sample. The Mn atoms suffer less local field effect due to their environments.

The doublet structure in the higher energy line shows the atomic characteristic of the Mn atom due to the fact that the Mn–Mn distance is larger than the separation of the Co atoms, and is attributed by Wang *et al.* (2005b) to the interplay of (1) the exchange and Coulomb interactions between core holes and unpaired valence electrons, and (2) the hybridization between 3d orbitals and surrounding states. In a thicker film, the smearing out of the structure according to Wang *et al.* is due to the reduction of orbital contributions as the thickness of the film increases. They also attributed the absence of multiple structures appearing in Sn-based Heusler alloys to more metallic Co d-states in Co_2MnSi.

The XMCD spectra are the normalized difference of the left- and right-handed polarized X-rays. They probe the exchange splitting and spin–orbit coupling of both initial core and final valence states.

A summary of the features in Fig. 3.17 is as follows:

- The structures in XMCD are in general 0.5 eV lower than the corresponding peaks of XAS.
- For the Co atom, the L_3 line is approximately a factor of two stronger than the L_2 line.
- For the Mn atom, the L_3 line is more than a factor of three stronger than the L_2 line and about a factor of two stronger than the corresponding Co line.
- Total intensities deduced from the sum rule correspond to 2.24 unoccupied states for the Co atom and 4.52 for the Mn atom.

For the Mn atom, the large negative dichroism within the L_3 region (lower panel in Fig. 3.17) is attributed to the large separation between the Mn atoms and smaller coordination numbers, resulting in weak

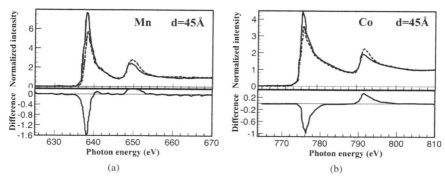

Fig. 3.17.   XMCD Co-L$_{2,3}$ lines (first and second peaks) of the Mn (a) and Co (b) atoms from a Co$_2$MnSi film of 45 Å thick (Wang *et al.*, 2005b).

Mn–Mn magnetic coupling. The Mn atom, consequently, exhibits a large magnetic moment. With 45 Å thick film, the sum rules provide information about the spin moments: $1.04\,\mu_B$ per Co atom, comparing well with the predicted value of $1.06\,\mu_B$. However, one must bear in mind that the sum rules underestimate both the effective spin moment and effective orbital moment.

Magneto-optical Kerr effect (MOKE)   Wang *et al.* (2005b) carried out MOKE measurements on film samples of Co$_2$MnSi at 70 K. The behavior of the Kerr rotation in the remanence along the $[1\bar{1}0]$ easy axis as a function of the thickness of the films is shown in Fig. 3.18. The results are summarized as follows:

- With thickness less than two bilayers, there is no magnetization. This could be due to magnetically inert layers or small islands preventing magnetic ordering.
- With thickness greater than two bilayers and an applied field of 1 kG, there is a nonzero Kerr signal. Beyond four bilayers, the behavior is close to linear.
- In-plane uniaxial magnetic anisotropy with the easy axis oriented along $[1\bar{1}0]$ is detected when the thickness increases beyond three bilayers. This is the critical layer thickness for the onset of FM long-range order.

The anisotropy in the plane of the film has been examined by measuring the Kerr rotation in a 60 Å film under a magnetic field with different angles between the easy $[1\bar{1}0]$ and hard $[110]$ axes. The geometry and results are

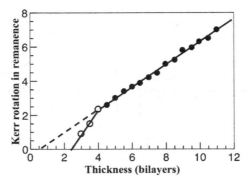

Fig. 3.18.   Thickness dependence of longitudinal Kerr rotation measured at 70 K for Co$_2$MnSi films. The open dots are the results for the films thinner than four bilayers (the superparamagnetic phase) in an applied field of 1 kG. The full dots show the saturated Kerr intensity of the FM phase (above four bilayers), which is equivalent to the intensity in remanence in this case (Wang *et al.*, 2005b).

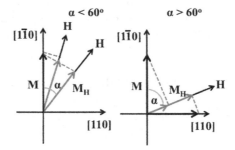

Fig. 3.19.   Geometric diagram of the magnetization and applied magnetic field (Wang *et al.*, 2005b).

shown in Fig. 3.19 and Fig. 3.20. Experiments were carried out at 80 K. The important results shown in Fig. 3.20 are:

- The appearance of sub-loops begins with magnetic field angle α greater than 60° with respect to the easy axis.
- For 65° < α < 75°, there are three loops, including a central loop.
- For $\alpha = 90°$, the three loops coalesce into one loop.

The appearance of sub-loops is explained by the applied magnetic field having enough strength to magnetize the sample along the hard axis. The origin of the anisotropy is unclear.

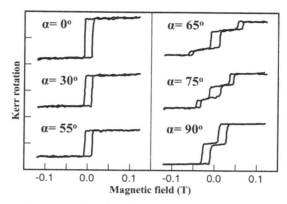

Fig. 3.20. Kerr rotation as a function of magnetic field applied in the plane of a 60 Å film (Wang *et al.*, 2005b).

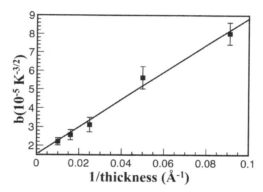

Fig. 3.21. Average spin-wave parameter b as a function of the inverse film thickness of the Co_2MnSi films grown on GaAs(001). The solid line is a linear fit to the experimental values (Wang *et al.*, 2005a).

The temperature dependence of Kerr rotations in different film thicknesses suggests that the magnetization can be well described by the usual Bloch formula, in particular for $T < 0.6T_C$,

$$M(T) = M(0)(1 - bT^{3/2}), \qquad (3.11)$$

where $M(0)$ and b depend on the thickness of the films. In Fig. 3.21, b is shown to be proportional to the inverse of the film thickness (Wang *et al.*, 2005a).

Physically, this formula describes the spin-wave excitations of an FM sample at finite temperature. The linear dependence shown in

Fig. 3.21 suggests that the film thickness decreases with the magnetization contributed by the surface and interface.

Spin-resolved photoemission spectroscopy (SRPES)   The SRPES spectra discussed below were obtained by Wang *et al.* (2005b) using the UE56/2-PGM2 beamline at BESSY in Berlin. The film samples were magnetized in-plane along the [$1\bar{1}0$] direction and have thicknesses ranging from 17 Å to 45 Å. The resolution of experiment is of the order of 0.5 eV. The results of spin spectra and polarization measurements are given in Fig. 3.22. The important features include:

- The overall intensity of the majority-spin spectrum is higher than that of the minority spin.
- There is a structure at the binding energy of 0.9 eV which is conjectured to arise from a mixture of metallic Co and Mn components.
- The spin polarization at Fermi level $P = (I_\uparrow - I_\downarrow)/(I_\uparrow + I_\downarrow)$ were calculated. The values at RT were 10% and 18% for the 17 Å and 45 Å samples, respectively. By correlating the temperature dependence of P to that of the magnetization, the extrapolated P at $T = 0$ is found to be 12%, independent of film thickness. This is quite low compared to the theoretical polarization of 100%.

Point-contact Andreev reflection (PCAR)   Singh *et al.* (2004b) carried out PCAR measurements on the Co_2MnSi thin films. The samples were grown by the dc magnetron co-sputtering method on an array of *a*-plane sapphire substrates. The base pressure was 2.67×10^{-9} mbar. The deposition rate was 0.10 nm/s. The substrate temperature ranged from 545 K

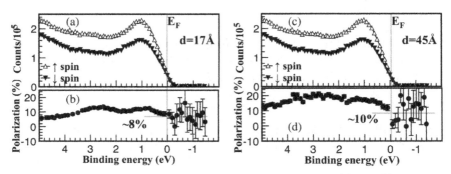

Fig. 3.22.   SRPES of two Co_2MnSi thin films measured at RT. (a) and (c) are spin-resolved spectra. (b) and (d) are spin polarization (Wang *et al.*, 2005a).

to 715 K. An optimized stoichiometric film of $L2_1$ structure was grown at 715 K with 400 nm thickness. However, it was in polycrystalline form with strong (110) texture. The P value measured with PCAR method was 56% at 4.2 K.

3.6.2.3. *Transport properties*

Temperature dependence of resistivity   Measurements of resistivity ρ as a function of T reveal the operative scattering mechanisms. For example, if ρ is proportional to T^2, the scattering may be electron–magnon or electron–electron in nature. A common method for measuring resistivity is the four-point DC method. For the stoichiometric films of 400 nm thick Co_2MnSi grown at various T_{sub}, the temperature of the a-plane sapphire substrate, Singh *et al.* (2004b) measured ρ at 4.2 K as a function of T_{sub}. As demonstrated in Fig. 3.23, the results show two distinct linear behaviors in different regions of T_{sub} divided at 570 K. The decrease of ρ is due to the increase of grain sizes.

It is more interesting to examine ρ as a function of T for a thin-film sample. Singh *et al.* (2004b) carried out these types of measurements on a sample grown at T_{sub} of 715 K. For 295 K $> T >$ 100 K, ρ decreases linearly with T, indicating electron–phonon scattering. For $T <$ 100 K, $\rho(T)$ fits well with $T^2 + T^{9/2}$. These authors attributed the second contribution to the electron–two–magnon scattering. They suggested that the T^2 dependence originated from either the electron–one–magnon or electron–electron scattering.

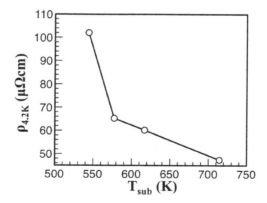

Fig. 3.23. ρ as a function of T_{sub} of a Co_2MnSi film (Singh *et al.*, 2004b).

Magnetic tunnel junctions (MTJs)   The quantity of interest in MTJs is the TMR. The basic principle was discussed in Chapter 2. The junctions have numerous device applications. They are essential components of sensors and memory devices.

Since fabricating MTJs is technologically important and an early theory predicted that disorder, such as antisites, can destroy half-metallic properties of Co_2MnSi (Ishida *et al.*, 1998), it is important to control precisely the microstructure of the sample. In the following, we discuss methods of growing films and junctions.

Kämmerer *et al.* (2003) grew Co_2MnSi films onto a vanadium(V)-buffer layer which can assist in (110) texture formation at RT using combined dc and radio frequency magnetron sputtering. They also found that it is effective to use V as the seed layer. However, the samples show low saturation magnetization. These authors (Kämmerer *et al.*, 2004) therefore suggested the following strategy for successful growth of Co_2MnSi films for MTJs.

- Prepare a magnetically optimized layer of Co_2MnSi through the evolution magnetic moment as a function of annealing temperature on a V-buffer layer.
- Grow an oxide layer, such as AlO_x, as a barrier by oxidizing an Al layer.
- Grow the upper magnetic electrode.
- Grow the upper current lead.

In practice, details of each step are important. During the growth, the base pressure was 1×10^{-7} mbar. In order to have a large TMR, it is crucial to have clean interfaces between Co_2MnSi and AlO_x and between AlO_x and $Co_{70}Fe_{30}$ — the upper magnetic electrode. This requires the whole growth process to take place in a vacuum chamber without any interruption.

The steps for growing MTJs having SiO_2 (substrate)/ V (42 nm)/ Co_2MnSi(100 nm)/ AlO_x(1.8 nm)/ $Co_{70}Fe_{30}$ (5.1 nm)/ Mn_{83}/Ir_{17} (10 nm) are:

- Use dc MS to grow the V buffer, Co_2MnSi, and Al layers.
- Oxidize the Al layer in pure oxygen plasma for 150 s.
- Anneal for 40 minutes at temperatures between 400 and 500 °C.

The last step is an important one. It can result in ferromagnetically textured Co_2MnSi with minimal disorder and homogenize the AlO_x barrier. It is tested by monitoring the magnetic moment/formula unit, M, of the FM layer. After annealing for one hour at a temperature above 250 °C,

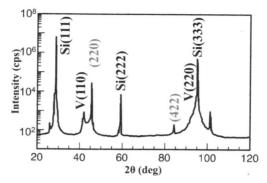

Fig. 3.24. X-ray diffraction pattern of a 100 nm Co_2MnSi film on a 42 nm V buffer layer taken after an annealing process of one hour at a temperature above 250 °C; the (hkl)-peaks belonging to Co_2MnSi are indicated in gray (Kämmerer *et al.*, 2004).

the value of M is more than $4.0\,\mu_B$. At 375 °C, M reaches $4.7\,\mu_B$. The annealing helps the (110) texture formation in the Co_2MnSi layer as shown in the X-ray diffraction pattern (Fig. 3.24). The existence of (110) texture was evidenced by the gray (220) line (Kämmerer *et al.*, 2004).

This line indicates that there is a periodic structure with periodicity twice or four times longer than the distance between adjacent (110) planes. When V is used, these two lines accompanied by the V(110) line and a shoulder of the V(220) line are clearly seen (Fig 3.24).

- Oxidize the top AlO_x layer for 50 seconds to clean the surface.
- Sputter $Co_{70}Fe_{30}$ by dc MS to form the upper magnetic electrode.
- Use radio frequency MS to put an AFM Mn_{83}/Ir_{17} layer on top.
- Apply dc MS again to form a Cu/Ta/Au multilayer lead.
- Impose an exchange bias between Mn_{83}/Ir_{17} and the upper FM $Co_{70}Ir_{30}$ electrode by annealing the multilayer structure for one hour at 275 °C under an external magnetic field with a strength of 100 mT.
- Pattern the multilayer material using optical lithography and ion beam etching to form 200 μm quadratic (four different layers) MTJs.

The next issue is the determination of the TMR. Two quantities were measured by Kämmerer *et al.* (2004) to determine the TMR of MTJs: the resistance R under a magnetic field and the difference of resistance ΔR when the magnetizations of the two electrodes are parallel vs. antiparallel. The ratio $\Delta R/R(H_{ext})$ of these two quantities is related to the TMR, where R is in general the resistance measured when the two electrodes have

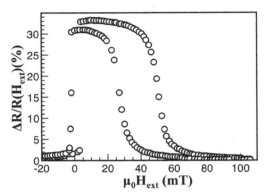

Fig. 3.25. $\Delta R/R(H_{ext})$ of an MTJ as a function of external magnetic field at RT (Kämmerer *et al.*, 2004).

parallel magnetization. The ratio plotted against the external field is shown in Fig. 3.25.

Figure 3.26 explains schematically the tunnel processes (Sakuraba *et al.*, 2006). Note that in a case to be discussed, both the sample and the upper electrode are made of the same materials. The upper panels are for the parallel configuration of magnetizations in the sample (at right) and upper electrode (at left). A bias potential $e\Delta V$ is applied such that a positive bias causes an electron to tunnel from the right to the left. The rectangular area indicates the tunneling barrier. The DOS showing the half-metallicity of the sample and upper electrode are plotted on both sides of the barrier. The gap in the minority-spin channel is denoted by E_G. δ_{CB} is the energy separation between the bottom of the conduction bands and E_F, and δ_{VB} is the energy separation between E_F and the top of the valence bands. When $e\Delta V = 0$, E_F is aligned on both sides. An electron in the majority-spin channel can tunnel from the sample to the upper electrode. When $e\Delta V > E_G$, E_F in the upper electrode is shifted down by $e\Delta V$. A new tunneling channel between minority-spin states is open as the result.

The lower panels are for the antiparallel configuration of magnetizations in the two electrodes. Schematically, it is represented by a reversal of spin channels in the sample. At $e\Delta V = 0$, an electron in the minority-spin ($\downarrow$) channel of the sample cannot tunnel to the left because of the insulating gap in the upper electrode, neither can a valence electron in a majority-spin ($\uparrow$) state due to the Pauli principle — the state in the upper electrode is filled. With finite bias, electrons in both channels can eventually

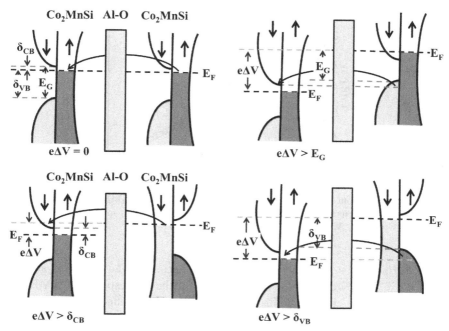

Fig. 3.26. Schematic of electron tunneling process at finite bias voltage in the $Co_2MnSi/Al–O/Co_2MnSi$ system for parallel (upper panels) and antiparallel (lower panels) configurations (Sakuraba *et al.*, 2006).

tunnel. The change of current at small bias voltage gives ΔR when a parallel configuration is switched to an antiparallel configuration.

A simple analysis given by Kämmerer *et al.* (2004) expresses TMR as a function of the difference between θ_{sample} and θ_{upper}, where θ_{sample} is the angle between magnetization of the sample (e.g., Co_2MnSi) and external magnetic field, and θ_{upper} is the angle between upper electrode, $Co_{70}Fe_{30}$ (Kämmerer *et al.*, 2004) or Co_2MnSi (Kämmerer *et al.*, 2003) and external field.

$$\text{TMR}(H_{\text{ext}}) = \frac{A_{\text{TMR}}(T)}{2}[1 - \cos(\theta_{\text{sample}} - \theta_{\text{upper}})], \tag{3.12}$$

$$M(H_{\text{ext}}) = \frac{M_{\text{sample}}\cos(\theta_{\text{sample}})}{[t_{\text{sample}}/(t_{\text{sample}} + t_{\text{upper}})]} + \frac{M_{\text{upper}}\cos(\theta_{\text{upper}})}{[t_{\text{sample}}/(t_{\text{sample}} + t_{\text{upper}})]}, \tag{3.13}$$

where A_{TMR} is the measured amplitude of the ratio $\Delta R/R(H_{\text{ext}})$. It is called the TMR-effect amplitude and is a function of T. The measured

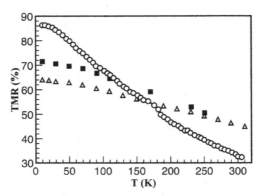

Fig. 3.27.   TMR-effect amplitude as a function of T in $Co_2MnSi/AlO_x/Co_{70}Fe_{30}$ $Mn_{83}Ir_{17}$ (open circles), $Mn_{83}Ir_{17}Co_{70}Fe_{30}/AlO_x/Co_{70}Fe_{30}$ (open triangles), and $Mn_{83}Ir_{17}Co_{70}Fe_{30}/AlO_x/Ni_{80}Fe_{20}$ (filled squares) (Kämmerer *et al.*, 2004).

$\Delta R/R(H_{ext})$ as a function of external magnetic field for MTJs is shown in Fig. 3.25. Using Eqs. (3.12) and (3.13), the corresponding TMR as a function of T is shown in Fig. 3.27. The TMR-effect amplitude at low temperature is 86%. More recently, Sakuraba *et al.* (2006) fabricated $Co_2MnSi/AlO/Co_2MnSi$ MTJs and determined the TMR as a function of T. At RT, the value of the TMR approaches 67%, while at 2 K the value is 570%.

We now discuss the relationship between TMR-effect and spin polarization P using Jullière's formula. The relation between TMR and polarizations of the Co_2MnSi (sample) and $Co_{70}Fe_{30}$ (electrode) at 10 K can be found through:

$$\text{TMR}(10K) = \frac{2P_{sample}(10K)P_{electrode}(10K)}{1 - P_{sample}(10K)P_{electrode}(10K)}. \tag{3.14}$$

One can deduce P_{sample} at 10 K from this relation when the measured TMR-effect is saturated. Using $P \sim 50\%$ for $Co_{70}Fe_{30}$ determined by Thomas (2003) and TMR-effect amplitude 86%, Eq. (3.14) gives $P \sim 61\%$ for the Co_2MnSi sample (Kämmerer *et al.*, 2004).

For MTJs fabricated by Sakuraba *et al.* (2006), the P value can also be determined from Jullière's formula. Using $P = 89\%$ for the Co_2MnSi electrode and TMR-effect amplitude of 570%, the P value of the Co_2MnSi sample is 83%. In addition, Schmalhorst *et al.* (2004) grew MTJs similar to Kämmerer *et al.* (2004) except they replaced $Co_{70}Fe_{30}$ by an AFM layer of $Mn_{83}Ir_{37}$. The P value for this sample is determined to be 86% at 10 K.

By comparing the values of P determined from different magnetic measurements, it is found that single crystalline Co_2MnSi films grown on GaAs(001) substrates show the smallest P values. The possible reasons for this discrepancy are: (i) there is another nonstoichiometric phase distributed inside the film, and (ii) a nonmagnetic metallic phase in the surface/interface regions could add an equal contribution to both spin components and thus reduce the polarization. Wang *et al.* (2005b) suggested that the possibility of the second reason is unlikely according to their Kerr effect as a function of thickness measurements which indicate that the surface/interface regions are magnetic.

To explain the 61% value of P which is substantially less than the predicted 100%, Schmalhorst *et al.* (2004) attributed the reduced value to spin scattering of tunneling electrons from the paramagnetic Mn ions and atomic disorder in the interface region. There is also the possibility of intermixing of Co and Mn, and Co atoms occupying Mn sites. As shown by Picozzi *et al.* (2004), with Mn sites occupied by the Co atoms, half-metallicity persists in Co_2MnSi. Neutron diffraction experiments (Raphael *et al.*, 2002) suggest that there are approximately 10–15% Co-Mn antisites existing in the sample, with Mn sites occupied by the Co atoms.

Based on the above discussions, it should be clear that the TMR of MTJs with the Co_2MnSi as the lower electrode (sample) depends critically on the choices of the seed layer and materials for the upper electrode.

3.6.3. *Co_2FeSi*

This Heusler alloy has the highest T_C (1100 K) and largest magnetic moment/formula unit (5.97 ± 0.05 μ_B) at 5 K known to date (Wurmehl *et al.*, 2006a). Since it has more attractive properties than the Co_2MnSi, a number of groups have focused on thin-film forms of Co_2FeSi. We shall discuss the growth of these forms, their characterizations, magnetic properties, and transport properties. Finally, magnetic tunnel junctions composed of Co_2FeSi will be discussed.

3.6.3.1. *Growth in thin-film form*

The radio frequency sputtering method is one of the methods used to grow thin-film forms of Co_2FeSi on either MgO(100) or Al_2O_3 ($11\bar{2}0$) substrates (Schneider *et al.*, 2006). Another one is the MBE method (Hashimoto *et al.*,

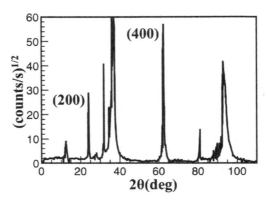

Fig. 3.28. X-ray Bragg reflection lines from a Co_2FeSi film grown on MgO(100) substrate (Schneider *et al.*, 2006).

2005) using GaAs(001) as the substrate. In the following, we discuss the radio frequency method. The target preparation was discussed in Chapter 2. The base pressure in the chamber was 5.0×10^{-8} mbar. The rate of deposition was 5 Å/s. The best samples were obtained with an Ar pressure of 1.0×10^{-2} mbar and T_{sub} of 700 °C. After the films were grown, a 4 nm Al layer was deposited on the top at 350 °C to prevent the oxidation.

3.6.3.2. *Characterizations*

Bragg reflection of X-rays has been used to characterize thin films (Schneider *et al.*, 2006). In Fig. 3.28, the Bragg lines of a Co_2FeSi film grown on MgO(100) substrate are shown. The (200) and (400) lines are sharp with widths of 0.3°. The results indicate that the growth in the out-of-plane direction is good. For the films grown on $Al_2O_3(11\bar{2}0)$ substrate, there are three different epitaxial domains. Only Co_2FeSi films grown on MgO are fully epitaxial (defect-free layers), as revealed by the LEED pattern (Fig. 3.29). The spots clearly show the four-fold symmetry.

3.6.3.3. *Magnetic properties*

XAS and XMCD experiments Magnetization was measured by XMCD using the UE56/1–SGM beamline at the BESSY-II Synchrotron light source (Schneider *et al.*, 2006). All measurements were under an external magnetic field of 16 kG to saturate the magnetization. The field was perpendicular ($\perp$) to the film surface. The measurements were carried out

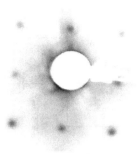

Fig. 3.29. The LEED pattern identifying epitaxial growth of Co_2FeSi grown on MgO (Schneider *et al.*, 2006). The primary electron beam energy is 318 eV.

at RT. The polarization of the X-ray was kept constant and the external field was reversed in direction. Total electron yield (TEY) measurements were carried out simultaneously.

The samples were 68 nm layers of Co_2FeSi grown on different substrates. Co $L_{2,3}$ and Fe $L_{2,3}$ lines were measured. For the XAS experiments, the total electron yield was measured in addition to the XAS. The TEY signal originates from the upper interface region within the electron escape depth. In order to collect all the escaping electrons, a conducting tube with a positive bias voltage (100 V) is used to shield the film under investigation. The photon flux of XAS transmitted through the sample is converted into a signal detected by a GaAs photodiode (PD) — the photodiode current I_{PD}. A transparent cap layer that is used to prevent the detection of any X-ray fluorescence from the sample protects the photodiode. The setup is schematically shown in Fig. 3.30 (Kallmayer *et al.*, 2007). I_{TEY} measures the intensity of TEY. The XMCD spectra were measured by flipping the magnetic field while keeping the polarization of the X-ray fixed.

Co_2FeSi (110) film on Al_2O_3 (11$\bar{2}$0) substrate Two Co $L_{2,3}$ lines determined by TEY are recorded in both XAS and XMCD experiments. The results are shown in Fig. 3.31. Assuming that the luminescence signal of the substrate $I_{lum}^{\pm}$ is proportional to the intensity of the transmitted X-ray and I_{ref} is the reference intensity from the bare substrate, $\mu^{\pm}$ are the absorption coefficients and can be calculated by $-\frac{1}{d}\ln[I_{lum}^{+}(h\upsilon)/I_{ref}(h\upsilon)]$, where d is the thickness of the film. $I_{TEY}(h\upsilon) = I_{lum}^{+}(h\upsilon) + I_{lum}^{-}(h\upsilon)$. Two peaks at 781 eV and 797.2 eV are observed in the XAS spectra. A smaller structure located at 3.5 eV above the 781 eV line is also observed,

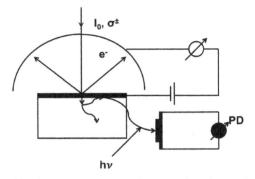

Fig. 3.30.   Schematic diagram of TEY measurements (Kallmayer *et al.*, 2007).

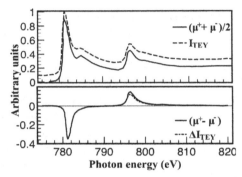

Fig. 3.31.   XAS and XMCD spectra of Co lines from a 68 nm Co_2FeSi film grown on a $Al_2O_3(11\bar{2}0)$ substrate (Schneider *et al.*, 2006).

which was noted as an indicator of the quality of the sample. It is noted that only the $L2_1$ structure shows this peak structure and it is absent in known disordered and selectively oxidized Heusler alloys. The corresponding structures in the XMCD spectra are shifted down by about 0.5 eV as in the case of Co_2MnSi.

For the Fe lines on the similar film sample, the two peaks in the XAS spectra are located at 711 and 724 eV, respectively (Fig. 3.32). Only a very weak shoulder at 715 eV is seen. The shifts of energies in the XMCD spectra are relatively small compared to the Co case. Using the sum rule with an assumed number of d-holes for each element, $n_d(Fe) = 3.4$ and $n_d(Co) = 2.5$ as reported for the pure elements (Wurmehl *et al.*, 2006b), magnetic moments of the Co and Fe atoms were determined. They are summarized in Table 3.17, where TMD indicates that the value

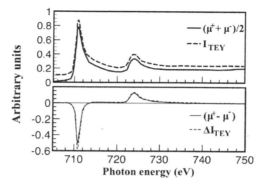

Fig. 3.32. XAS and XMCD spectra of Fe lines from a 68 nm Co$_2$FeSi film grown on a Al$_2$O$_3$(11$\bar{2}$0) substrate (Schneider *et al.*, 2006).

Table 3.17. Element-specific magnetic moments (μ_B) for Co and Fe in Co$_2$FeSi films grown on a Al$_2$O$_3$(11$\bar{2}$0) substrate.

Moment type	Co	Fe	Co$_2$FeSi
Spin (TEY)	1.13	2.47	4.73
Spin (TMD)	1.25	2.43	4.93
Orbit (TEY)	0.14	0.10	0.38
Orbit (TMD)	0.11	0.07	0.29
Sum (TEY)	1.27	2.57	5.11
Sum (TMD)	1.36	2.50	5.22
SQUID			4.8

was deduced from transmission data. The ratios of the orbital and spin moments for Co and Fe are 0.12 and 0.04, respectively, from the TEY results. The ratio for the Co atom (0.13) agrees well with that in crystalline Co (Carra *et al.*, 1993). However, for Fe, the present result compared to the crystalline value reported by Carra *et al.* is approximately 33 times smaller. One possible explanation is that moments of the neighboring Co atoms can orient in the opposite direction relative to that of the Fe atom. The sum of the spin and orbital momenta for the two measurements is larger than the result obtained by SQUID. The SQUID measurement was carried out at 300 K. The discrepancy was attributed to the assumed number of holes used in determining the magnetic moments from the X-ray data.

Co$_2$FeSi(100) film on MgO(100) substrate The results of XAS and XMCD measurements of Co and Fe lines from a 68 nm Co$_2$FeSi layer grown on

Table 3.18. Element-specific magnetic moments (μ_B) for Co and Fe in Co_2FeSi films grown on MgO(100) substrate.

Moment type	Co	Fe	Co_2FeSi
Spin (TEY)	1.07	2.46	4.60
Spin (TMD)	1.28	2.46	5.02
Orbit (TEY)	0.04	0.05	0.13
Orbit (TMD)	0.13	0.12	0.38
Sum (TEY)	1.11	2.51	4.73
Sum (TMD)	1.41	2.58	5.40
SQUID			4.8

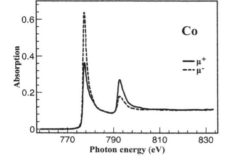 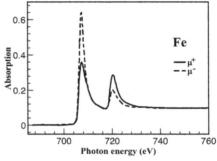

Fig. 3.33. XAS spectra of Co and Fe (Chen *et al.*, 1995).

MgO(100) substrate are similar to those shown in Figs. 3.31 and 3.32. The element-specific magnetic moments for the Co and Fe atoms deduced from the measurements and sum rule are given in Table 3.18. The orbital moments of the Co and Fe atoms deduced from the TEY spectra are three and two times smaller than those found on the Al_2O_3 ($11\bar{2}0$) substrate. For the transmission spectra, the results for both substrates agree reasonably well. The experimental results of Schneider *et al.* (2006) were compared to the XAS spectra of pure Co and Fe by Chen *et al.* (1995), as shown in Fig. 3.33. The peak positions shown in Figs. 3.31 and 3.32 agree well with those shown in Fig. 3.33.

The above results suggest that localized properties, such as the element-specific spin moment, are not affected by whether the sample is in thin-film form or bulk. Theoretical predictions of spin moments of the Co and Fe atoms based on LDA by Kandpal *et al.* (2006) are $2.3 \mu_B$ and $3.5 \mu_B$, respectively. The spin moment calculated with LDA+U (Kandpal *et al.*, 2006) of Co is $1.53\,\mu_B$ and that for Fe is $3.25\,\mu_B$. The magnetic moment is $6.0\,\mu_B$/unit cell. The last value was confirmed experimentally (Wurmehl *et al.*, 2005). The deduced element-specific spin moments given in Tables 3.17 and 3.18 are smaller than the theoretical values, in particular, the LDA results.

Magneto-optical Kerr effect (MOKE) As discussed previously in the case of Co_2MnSi, MOKE provides measurements of the anisotropy of a film. It is a first-order effect and is divided into:

- Polar MOKE (PMOKE), which is proportional to the out-of-plane magnetization.
- Longitudinal MOKE (LMOKE) M_L, which is related to the in-plane magnetization parallel to the plane of incidence light.
- Transverse MOKE (TMOKE) M_T, which provides information about the in-plane magnetization perpendicular to the plane of incidence light.

The relative orientations of the magnetization, external magnetic fields $\mathbf{H}_i$, and crystal axes of the film are shown in Fig. 3.34 (Hamrle *et al.*, 2007).

The MOKE loops in most Heusler alloy films are symmetric with respect to angle α (Fig. 3.34). As experiments have progressed, evidences have emerged that MOKE loops can be asymmetric. Postava *et al.* (2002)

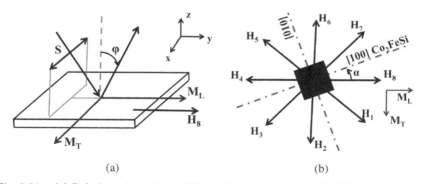

(a)   (b)

Fig. 3.34. (a) Relative orientations of the incident s-polarized light (S), magnetization in the film (M_L and M_T), and the external magnetic field $\mathbf{H_8}$. (b) Relative orientations of crystal axes, magnetization of the film, and the external magnetic field in different orientations $\mathbf{H_i}$. The electric field is perpendicular to the beam velocity (Hamrle *et al.*, 2007).

attributed the asymmetry to the quadratic behavior of MOKE — QMOKE, which arises from quadratic terms such as $M_L M_T$ and $M_L^{2-} M_T^2$.

Physically, the observation of the Kerr effect depends on both the spin–orbit and exchange interactions. Unlike PMOKE and LMOKE, where the magnetization $\mathbf{M}$ is parallel to the wave vector of the light, QMOKE requires $\mathbf{M}$ to be perpendicular to the plane of incidence.

Therefore, the first-order contribution of the spin–orbit interaction does not play a role. The second-order contribution of the spin–orbit interaction is smaller so QMOKE is a smaller effect. To show that QMOKE depends on the quadratic contributions of M_L and M_T, we outline the macroscopic derivations (Foner, 1956) as follows: the response function of a magnetic medium to the light can be described by the dielectric tensor,

$$\varepsilon_{ij} = \varepsilon_{ij}^{(0)} + K_{ijk} M_k + G_{ijkl} M_k M_l + \cdots , \tag{3.15}$$

where $\varepsilon_{ij}^{(0)}$ is the linear dielectric tensor, and K_{ijk} and G_{ijk} are the linear and nonlinear magneto-optical tensors, respectively. Repeated indices are used as the summation convention. The symmetries required by the Onsager relation give the following conditions:

$$\varepsilon_{ij}^{(0)}(\mathbf{M}) = \varepsilon_{ij}^{(0)}(-\mathbf{M}) = \varepsilon_0 \delta_{ij},$$

$$K_{ijk} = -K_{jik}, K_{iik} = 0, i \neq j \neq k, \tag{3.16}$$

$$G_{ijkl} = G_{jikl} = G_{jilk} = G_{ijlk},$$

where δ_{ij} is the Kronecker delta function. In a cubic crystal, the number of components of K and G can be further reduced to

$$K_{ijk} = K,$$

$$G_{iiii} = G_{11}, G_{iijj} = G_{12}, \tag{3.17}$$

$$G_{1212} = G_{1313} = G_{2323} = G_{44}.$$

With in-plane magnetization, the off-diagonal ε_{ij} are given as follows:

$$\varepsilon_{xy} = \varepsilon_{yx} = 2G_{44} M_L M_T,$$

$$\varepsilon_{xz} = -\varepsilon_{zx} = K M_L, \tag{3.18}$$

$$\varepsilon_{yz} = -\varepsilon_{zy} = -K M_T,$$

for $\alpha \neq 0$,

$$\varepsilon_{xy} = \varepsilon_{yx} = [2G_{44} + G_{\mathrm{map}}(1 - \cos(4\alpha))]$$

$$\times M_L M_T - G_{\mathrm{map}}\frac{\sin(4\alpha)}{4}(M_L^2 - M_T^2),$$

$$\varepsilon_{xz} = -\varepsilon_{zx} = KM_L,$$

$$\varepsilon_{yz} = -\varepsilon_{zy} = -KM_T, \tag{3.19}$$

where $G_{\mathrm{map}} = G_{11} - G_{12} - 2G_{44}$ and is called the magneto-optical anisotropy parameter. Using the well-established relations between Kerr rotations for the s polarization (the electric field of the incident light is parallel to the velocity of the incident light) and the dielectric tensor ε_{ij}, the Kerr s effect Θ_s can be expressed as:

$$\Theta_s = A_s \left[G_{44} + \frac{G_{\mathrm{map}}}{2}(1 - \cos(4\alpha)) + \frac{K^2}{\varepsilon_0} \right]$$

$$\times M_L M_T - A_s G_{\mathrm{map}} \sin(4\alpha)(M_L^2 - M_T^2) - B_s \frac{K}{K_L}, \tag{3.20}$$

where A_s and B_s are the weighting optical factors and are even and odd functions of angle φ (Fig. 3.34). A similar equation is for the Kerr p effect Θ_p with sign changes on A and B. Both of these constants depend on the incident angle φ.

The QMOKE measurements on the Co_2FeSi film at $\alpha = -22.5°$ were carried out by Hamrle *et al.* (2007). The sample was a 21 nm Co_2FeSi film on the MgO(100) substrate. A red laser of wavelength 670 nm with s polarization was used in the experiments. The laser spot diameter was approximately 300 μm. To determine the saturation signal of QMOKE, the Kerr angle was measured after the external field $\mathbf{H}_i$ was applied in the eight different directions (Fig. 3.34) with strengths sufficient to saturate the magnetization. These measurements can give rotations associated with $M_L M_T$ and $M_L^2 - M_T^2$ terms. In Fig. 3.35, QMOKE rotations with respect to the external field $\mathbf{H}_8 \parallel y$ and saturation are shown. The filled circle at $H = 0$ is the signal of QMOKE at saturation. The linear part arises from the stray field of the magnet on the optical components of the experimental setup. As the field increases from saturation, the signal of QMOKE decreases to zero as indicated by the arrows, one for the field in positive and the other in negative directions. At that point, the average values of $M_L M_T$, $\langle M_L M_T \rangle$, and $M_L^2 - M_T^2$, $\langle M_L^2 - M_T^2 \rangle$, over the

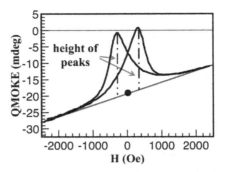

Fig. 3.35.   QMOKE Kerr rotation loops in a 21 nm Co_2FeSi film with $\varphi = 0.5°$ and $\alpha = -22.5°$. The external magnetic field is in $H_8 \parallel y$ direction. The filled circle at $H = 0$ is the QMOKE signal in saturation as determined by the 8-directional method for $\alpha = -22.5°$ (Hamrle *et al.*, 2007).

laser spot are equal to zero simultaneously. This indicates the presence of magnetic domains during the reversal process; otherwise the two quantities can not be equal to zero simultaneously. Furthermore, the height of the QMOKE signal peaks reaches 30 mdeg indicating that there is a large second order or even higher order spin–orbit coupling in the sample.

Spin-resolved photoemission spectroscopy (SRPES)   Schneider *et al.* (2006) also carried out SRPES experiments. In this section, we discuss some practical aspects. The film samples were placed in a UHV with base pressure below 10×10^{-10} mbar. The Al capping layer on top of the film was removed by sputtering with Ar^+ ions of energy 500 eV. The surface was then subjected to repeated cycles of sputtering and prolonged annealing at 570 K. The films were magnetized by an external in-plane magnetic field. The experiments were carried out at 300 K.

The incident light was an *s*-polarized 4-th harmonic narrow band with energy centered at 5.9 eV in the form of a pulse generated by a Ti:sapphire oscillator. The light was incident on the sample at a 45° angle with respect to the surface normal.

The photoemitted electrons were analyzed in a normal emission geometry by a commercial cylindrical sector analyzer equipped with an additional spin detector based on spin-polarized LEED. The energy resolution and acceptance angle of the analyzer are 150 meV and ±13°, respectively.

The SRPES spectra for films grown on Al_2O_3 and MgO substrates are shown in Fig. 3.36. For comparison, shown in Fig. 3.37 is the bulk DOS

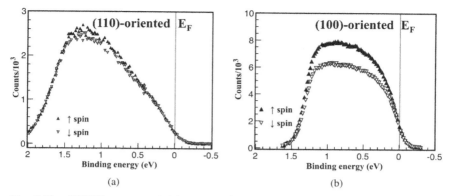

Fig. 3.36. SRPES spectra of (a) Co_2FeSi/Al_2O_3 and (b) Co_2FeSi/MgO (Schneider *et al.*, 2006).

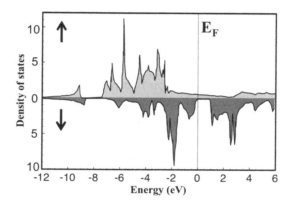

Fig. 3.37. Calculated DOS of bulk Co_2FeSi (Wurmehl *et al.*, 2005).

of Co_2FeSi from the LDA+U band structure calculations (Wurmehl *et al.*, 2005). The SRPES spectra show relatively smooth variation near E_F and broad asymmetric peaks centered at 1.3 eV in both spin channels.

The theory shows a wide peak in the ↓ spin channel near −1.0 eV. There is some agreement between SPRES spectra and the calculated DOS for minority-spin states in the energy range between −1.5 eV to E_F (E_F is set to be zero).

Experimentally, the majority-spin states have a slightly higher intensity over the whole energy range of measurements for both films. Therefore, the spin polarization is positive. The maximum spin polarization is 4% for the Co_2FeSi/Al_2O_3 film and 12% for the Co_2FeSi/MgO film. $P(E_F)$ is 0 for the former and 6% for the latter. The reduction of the spin polarization is not confined to the top surface layer. It extends 4–6 nm into the sample for the Co_2FeSi/MgO film. Compared to XAS and XMCD, which measure the integrated spectra, SRPES probes states near E_F in a very thin region of the surface. These spectra should therefore be more sensitive to disorder in the surface region.

Magnetization Schneider *et al.* (2006) measured the saturation magnetization for Co_2FeSi films grown on MgO and Al_2O_3 substrates. The extrapolated value to $T = 0$ K is $5.0\,\mu_B$/formula unit for those films on Al_2O_3. There is a uniaxial anisotropy in films grown on Al_2O_3 with the easy axis along the [110] direction. Films deposited on the MgO substrate show no significant anisotropy.

There is a $T^{3/2}$ dependence in $M(T)$ for Co_2FeSi grown on MgO. The result is shown in Fig. 3.38. This dependence indicates spin-wave excitations in these films.

Curie temperature T_C As we have mentioned previously, Co_2FeSi has the highest T_C among all Heusler alloys. Wurmehl *et al.* (2005) measured the high-temperature magnetization of Co_2FeSi using the VSM method (Fig. 3.39). The measurements were carried out under a magnetic field of 1 kG. The quantity determined is the inverse of the susceptibility $1/\chi(T)$ for $T > T_C$. This is based on the Curie–Weiss law:

$$\chi(T) = \frac{C}{T - T_C}. \tag{3.21}$$

The value of T_C is 1100 ± 20 K (Wurmehl *et al.*, 2005).

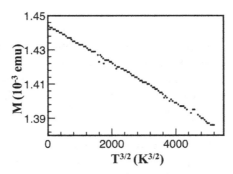

Fig. 3.38. $M(T)$ as a function of temperature for a Co_2FeSi film grown on the MgO substrate (Schneider *et al.*, 2006).

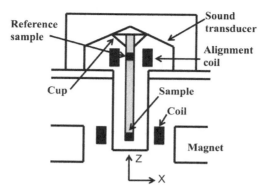

Fig. 3.39. A schematic setup of VSM. The coil senses the signal proportional to χ. The cup transmits the sound wave to the sample (Foner, 1956).

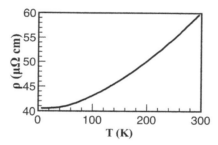

Fig. 3.40. Resistivity as a function of temperature in a typical Co_2FeSi film (Schneider *et al.*, 2006).

3.6.3.4. *Transport properties*

Temperature dependence of resistivity The van der Pauw four-probe method was used by Schneider *et al.* (2006) to measure the resistivity ρ. The temperature range was $0 < T < 300\,\text{K}$. Results for a typical film of Co_2FeSi are shown in Fig. 3.40. The residual resistivity ρ_o defined at $T = 4.0\,\text{K}$ is $40\,\mu\Omega\text{cm}$. The corresponding ratio between the resistance at $300\,\text{K}$ and $4.0\,\text{K}$ is 1.5. These values fall in the range of other Heusler alloys in thin-film form. The key feature of $\rho(T)$ is the exponent. The resistivity curve shown in Fig. 3.40 can be fitted to:

- $\rho(T) \sim T^{7/2}$, $T < 70\,\text{K}$.
- $(\rho - \rho_o) \sim T^{1.65}$, $T > 70\,\text{K}$.

These results do not indicate that the dominant scattering mechanism is the one-magnon process. Other processes, such as incoherent scattering and the scattering between s- and d-states, are possible causes of exponents deviating from 2.0. It is also possible to get better agreement with a one-magnon process if the shape of the Fermi surface is taken into account.

Magnetoresistance This quantity measures the current–voltage response of a sample under an external magnetic field **H**. The resistance can have two components:

- $\rho_{\parallel}$, the resistivity parallel to the magnetic field.
- $\rho_{\perp}$, the resistivity perpendicular to the magnetic field.

The measured results by Schneider *et al.* (2006) are shown in Fig. 3.41. $\rho_{\perp}$ (the top curve) is larger than $\rho_{\parallel}$ (the bottom curve). This is attributed to the spin–orbit interaction. At $H = 0$, the spontaneous anisotropy can be determined from the expression,

$$\frac{(\rho_{\parallel} - \rho_{\perp})}{\frac{1}{3}(\rho_{\parallel} + 2\rho_{\perp})}. \tag{3.22}$$

The ratio is found to be –0.8%, which is considered small. At large field $(H > 15\,\mathrm{kG})$, both ρs show linear behavior with different slopes that are functions of temperature. Similar behaviors have been observed in conventional ferromagnets. The resistivity has been attributed to spin-flip scattering processes.

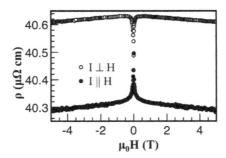

Fig. 3.41.   Magnetoresistivity of a thin-film Co_2FeSi at $4\,\mathrm{K}$ (Schneider *et al.*, 2006).

Tunnel magnetoresistance (TMR)   The TMR of Co_2FeSi films grown on MgO has been measured with the film as one of the electrodes (Niculescu *et al.*, 1977). At RT, TMR is 41%. The value increases to 60% at 5K when the film grown at 473 K is post annealed at 573 K. The results show that TMR is highly sensitive to the sample preparation.

Chapter 4

Half-Metallic Oxides

4.1. Introduction

Among the various oxides, a few are now known to show half-metallic properties. They are:

- chromium dioxide (CrO_2),
- magnetite (Fe_3O_4),
- manganites, $La_{2/3}Sr_{1/3}MnO_3$ (LSMO) and $La_{1-x}Ba_xMnO$.

The last of these emerged from the study of high T_C superconductors. All three are appealing for spintronic applications because they are predicted to have 100% spin polarization at E_F. Experimentally, CrO_2 exhibits P up to 97% at 4 K (Ji et $al.$, 2001). These materials are expected to show large TMR when used to form electrodes sandwiching a thin insulating region, and so are good candidates for spintronic devices such as spin valves and sensors.

The above three oxides are of comparable complexity to the Heusler alloys with respect to growth and structure. In this chapter, we shall discuss each one separately, considering in detail the growth, characterization, and magnetic and electronic properties for each.

4.2. CrO_2

Experimental studies of CrO_2 go back to the late 1960s (Stoffel, 1969). Schwarz (1986) was the first to predict CrO_2 to be a half-metallic ferromagnet. This study was motivated by the potential for using this material to make read-write magneto-optical memory devices.

Table 4.1.   Positions of Cr and O atoms
in rutile structure expressed in reduced
coordinates, where δ is 0.3053.

Atom	Coordinates (x,y,z)
Cr	(0,0,0)
Cr	(0.5, 0.5, 0.5)
O	(δ, δ, 0)
O	($-\delta$, $-\delta$, 0)
O	($0.5 + \delta$, $0.5 - \delta$, 0.5)
O	($0.5 - \delta$, $0.5 + \delta$, 0.5)

4.2.1. *Structure*

CrO_2 has the simplest structure among the three oxides discussed above. It crystallizes in the rutile (TiO_2) structure (Fig. 1.3).

The space group is D_{4h}^{14} ($P4_2/mnm$), which is a nonsymmorphic group. There are two formula units per unit cell. The Cr atoms form a body-centered tetragonal lattice, as shown in Fig. 1.3. Each of the Cr atoms is surrounded by an octahedron of O atoms. The orientations of the two octahedra differ by a 90° rotation about the z-axis (parallel to the shorter lattice vector, c). The positions of the Cr and O atoms are given in Table 4.1. Denoting the longer lattice vector by a in the x-y plane, the c/a ratio is 0.65958 (Schwarz, 1986). The lattice constant, a, was determined by Thamer *et al.* (1957) to be 4.42 Å. Porta *et al.* (1972) confirmed these results.

4.2.2. *Growth*

Cr forms many competing oxides, namely Cr_3O_4, Cr_2O_5, CrO_3, and Cr_2O_3. Crystalline CrO_2 is the only oxide which is FM at RT. But it is metastable at atmospheric pressure and is easily decomposed to Cr_2O_3 with heat. Therefore, it can be difficult to grow. The successful growth techniques can be classified into two categories: the "high-pressure" and "low-pressure" techniques. Thermal decomposition (Hwang and Cheong, 1997; Ranno *et al.*, 1997) is the high-pressure technique. Chemical vapor deposition (CVD) (Kämper *et al.*, 1987; Li *et al.*, 1999; Gupta *et al.*, 2000; Anguelouch *et al.*, 2001; Ivanov *et al.*, 2001) is the low-pressure technique. More recently, PLD has been used to grow CrO_2 (Shima *et al.*, 2002). PLD (discussed in Chapter 3) is a low-pressure technique. In the following, we shall focus on the thermal decomposition and CVD techniques.

4.2.2.1. *Thermal decomposition*

This technique was successfully used by Ranno *et al.* (1997) to grow powder and film samples of CrO_2. For the powder growth, CrO_3 was decomposed in a closed vessel with different oxygen pressures under a controlled temperature. The reaction can be described by the chemical formulas:

$$CrO_3 \rightarrow CrO_2 + \frac{1}{2}O_2 \tag{4.1}$$

and

$$2CrO_3 \rightarrow Cr_2O_3 + \frac{3}{2}O_2. \tag{4.2}$$

One must bear in mind that the end product can be Cr_2O_3. Ranno *et al.* used the thermopiezic analysis (TPA) to record the decomposition temperature and oxygen release. An instructive result of the TPA shows the range of pressure and temperature for obtaining CrO_2 from a powdered sample of CrO_3 as given in Fig. 4.1. CrO_3 decomposes at 250 °C at ambient pressure. Between 300 and 400 °C is the region forming the mixture of CrO_2 and Cr_2O_5 with a trace of Cr_2O_3. CrO_2 powder decomposes at 450 °C. To grow thin-film forms of CrO_2, a single crystal of TiO_2 or Al_2O_3, with an area of a few mm^2 and thickness of 0.5 mm, serves as the substrate, and powdered CrO_3 as the precursor in a brass reactor with Al seal. The reactor was put into a furnace and heated at 0.5 °C/min from RT. Before bringing back to RT at a rapid rate, the reactor was held at 425 °C for about an hour.

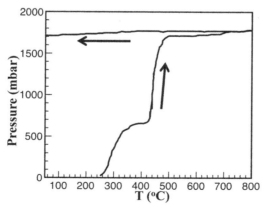

Fig. 4.1. The oxygen pressure as a function of temperature recorded by TPA for an 18 mg CrO_3 sample under heating at 0.5 °C/min (Ranno *et al.*, 1997).

4.2.2.2. *Chemical vapor deposition (CVD)*

The methods used to grow CrO_2 using CVD differ from one another mainly in precursor. Most have used the decomposition of CrO_3 in a closed reactor onto heated substrates of RuO_2, TiO_2, or Al_2O_3. In a typical case, the reaction tube has a two-zone furnace (Li *et al.*, 1999). Powdered CrO_3 was loaded in the first zone and kept at $260\,°C$. Oxygen carriers flow from the first to the second zone at the rate of $0.5\,cc/min$ at atmospheric pressure. The substrate was in the second zone and was heated to $390-400\,°C$.

The grown films were put into a UHV chamber and the surfaces were cleaned by soft sputter cycles of 30 seconds with a defocused $500\,eV$ Ne ion beam at grazing angle of incidence. This method was used because the surface of CrO_2 can easily transform into Cr_2O_3 at $200\,°C$. The samples are in polycrystalline form.

Gupta *et al.* (2000) grew finer-grain polycrystalline CrO_2 films on a seed layer of TiO_2. The samples grown on Al_2O_3 are highly oriented due to the six-fold symmetry of the substrate with multiple grains. A year earlier, Gupta *et al.* (1999) reported the growth of CrO_2 thin films by CVD with selective-area growth in specific regions, followed by lateral epitaxial overgrowth over masked regions. The substrate was a single-crystal TiO_2. The selective-area and lateral overgrowth was accomplished using a prepatterned SiO_2 mask.

Anguelouch *et al.* (2001) used a liquid precursor CrO_2Cl_2. It was placed in a bubbler and kept at $0\,°C$. Their reactor was a quartz tube inside a furnace. The tube was kept at $400\,°C$. The substrate was TiO_2 oriented in the (100) plane. It was placed on a tilted glass susceptor after the use of organic solvents and a dilute hydrofluoric acid to clean the tube. With oxygen gas flowing at $40\,cc/min$, these authors grew thin-film forms of CrO_2 at a deposition rate of $72\,Å/min$.

Ivanov *et al.* (2001) provided some insight into the growth of CrO_2 from CrO. They realized that in the CVD technique the oxygen gas serves as the mechanical carrier of the CrO_3 and is not crucial to the growth. They tried to use MBE with a CrO_3 beam but failed to get any CrO_2. The work of Norby *et al.* (1991) identified that one of the intermediates is Cr_8O_{21}. This led Ivanov *et al.* to prepare Cr_8O_{21} as the sole precursor by heating powdered CrO_3 to $250\,°C$ for a period of eight hours. With this precursor, they finally grew high-quality epitaxial CrO_2 layers on $Al_2O_3(0001)$ and $TiO_2(110)$ substrates. They also pointed out that further heating of CrO_3 to $330\,°C$ for twelve hours produced Cr_2O_5. This sample cannot be used as a precursor to grow CrO_2. Therefore, one must be extremely careful to control the temperature when growing CrO_2.

4.2.3. *Characterization*

There are three techniques used to characterize the quality of CrO_2 in thin-film form: RHEED (Li *et al.*, 1999), XRD (Gupta *et al.*, 1999), and AFM (Anguelouch *et al.*, 2001).

4.2.3.1. *Reflection high-energy electron diffraction (RHEED)*

Li *et al.* (1999) aligned the electron beam along orthogonal zone axes and determined the RHEED pattern *ex situ* for TiO_2 and Al_2O_3 substrates,

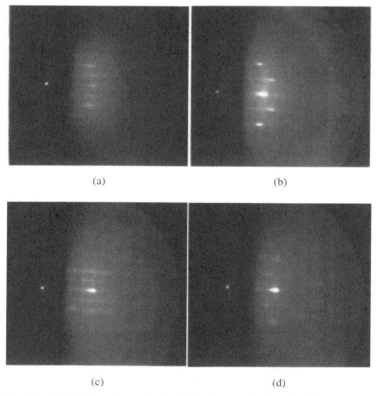

(a) (b)

(c) (d)

Fig. 4.2. The RHEED patters obtained by Li *et al.* (1999). (a) On $TiO_2(100)$, beam $\langle 001 \rangle$, (b) on $TiO_2(100)$, beam $\langle 110 \rangle$, (c) on $Al_2O_3(0001)$, beam $\langle 1\bar{1}00 \rangle$, (d) on $Al_2O_3\langle 0001 \rangle$, beam $\langle 11\bar{2}0 \rangle$.

respectively. The results are shown schematically in Figs. 4.2(a) and 4.2(b). The patterns are sharp and streaky on TiO_2, suggesting the surface is well-ordered and flat. The pattern shown in Fig. 4.2(c) is formed by small spots even though it is still sharp. The pattern in Fig. 4.2(d) shows hexagonal spots. These patterns suggest the films were grown in three dimensions. The patterns for different beam directions in the growth on a TiO_2 substrate show that the film axes are aligned with the axes of the substrate. However, the beam directions indicate there are multiple domains with the b-axis of CrO_2 aligned along the equivalent directions of $\langle 11\bar{2}0 \rangle$ on Al_2O_3.

4.2.3.2. *X-ray diffraction (XRD)*

By using the selective-area and lateral growth techniques, Gupta *et al.* (1999) obtained epitaxial growth of CrO_2. Three surfaces of the substrate are used: (100), (110), and (001). The normal X-ray intensity as a function of 2θ of those three cases is given in Fig. 4.3. The first peak in each case is associated with TiO_2 and the second peak with CrO_2, because the lattice constants of the sample and substrate are different. The lines for TiO_2 are very narrow for all three directions, and are reproducible, indicating that substrate surfaces were all properly cleaned. The lattice-constant mismatches are large in the three orientations. For (100) and (010) orientations, it is -3.79%, and for (001) it is -1.48%. Therefore, there is a tensile strain in the in-plane direction and a compressive strain in the

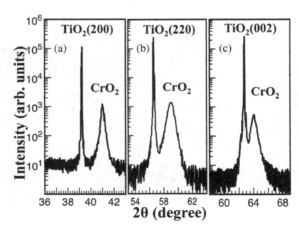

Fig. 4.3.   X-ray intensity as a function of 2θ (degree) with the substrate surfaces: (a) (200), (b) (220), and (c) (002) (Gupta *et al.*, 1999).

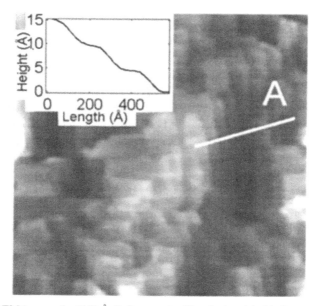

Fig. 4.4. AFM image of a 1050 Å CrO_2 sample. The distinguishable lines correspond to the atomic planes of CrO_2. The inset shows a sample cross section of the region marked a with 4.4 Å steps (Anguelouch *et al.*, 2001).

perpendicular direction. The peaks in case (c) show the largest shift to the higher angle. The largest full width at half maximum for the rocking curve[1] of the second peak in case (b) is 0.4°, which is considered to be narrow.

4.2.3.3. *Atomic force microscopy (AFM)*

The AFM image of 1050 Å CrO_2 sample obtained by Anguelouch *et al.* (2001) is shown in Fig. 4.4. The distinguishable lines correspond to atomic planes of CrO_2. The root-mean-square roughness in an area of $1.0\,\mu m^2$ is 4.6 Å. The height of each step is either 4.4 Å or 8.8 Å. These steps correspond to the height of one or two planes. The films are atomically smooth.

---

[1]The rocking curve provides information about compositional and thickness variations. In addition, lattice-constant difference between a substrate and a sample can be determined within 10 parts per million from the peak separation and the ratio of integrated intensities of the two peaks of the rocking curve.

4.2.4. *Transport properties*

Since applications of MR are common for many oxides besides CrO_2, and a technique — the point-contact method — has been developed for powdered forms of oxides, we shall discuss first this technique and the associated powder magnetoresistance (PMR). We are particularly interested in the half-metallic properties of CrO_2. The experimental determination of these properties will be discussed. Finally, the electronic properties of this compound will be discussed based on photoemission experiments and theoretical grounds.

4.2.4.1. *Point contacts and powder magnetoresistance (PMR)*

Many oxides are grown in powdered form. One of the more effective experimental techniques for determining the PMR of the oxides, developed by Coey *et al.* (2002), is the point-contact method:

- Definition: A point contact is a small region of atomic scale continuity between two crystallites (Fig. 4.5). It is possible to have a grain boundary at the contact. This last property differs from a nanoconstriction in which there is crystallographic coherence and no grain boundary.
- Both the point contact and nanoconstriction can have domain walls which can be pinned at the contact.
- The tunneling process to get MR is through the point contacts.

Let δ be the wall width and λ be the mean free path (MFP). If $\lambda > \delta$, then the mechanism for the transport is ballistic. The transmission coefficient T_b for an HM oxide under low bias and zero temperature (Coey *et al.*, 2002) is

$$T_b = \cos^2\left(\frac{\theta_{12}}{2}\right), \tag{4.3}$$

where θ_{12} is the angle of misalignment of the magnetization in the domains on each side of the contact. For $\lambda \ll \delta$, the transport is diffusive and the transmission coefficient T_d (Coey *et al.*, 2002) is

$$T_d = \frac{1}{2}\left[\cos\left(\frac{\theta_{12}}{2\nu}\right)\right]^{\nu-1}, \tag{4.4}$$

where $\nu = \delta/\lambda$. The MR is

$$\text{MR} \sim (1 - T_{b/d}). \tag{4.5}$$

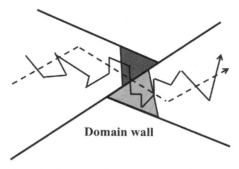

Domain wall

Fig. 4.5. Schematic of a point 3 contact (Coey *et al.*, 2002). The shaded triangular areas are two magnetic domains. Their magnetizations are oriented relative to each other by θ_{12} (not indicated). The dashed lines with arrows indicate a path of ballistic transport while the solid lines show the diffusive transport from the left to the right.

Note that $\theta_{12} = \pi$ gives the largest MR.

$$\text{MR} = \frac{R(0) - R(H)}{R(0) + R(H)} = \frac{G(H) - G(0)}{G(H) + G(0)}, \qquad (4.6)$$

where G is the conductance. If the samples are in pressed powder form and the magnetic moments in the two domains are not completely aligned along the external magnetic field, then

$$\text{PMR} = \frac{m^2}{(1 + m^2)}, \qquad (4.7)$$

where $m^2 = \langle cos\theta_{12}\rangle^2$.

Setup The setup for point-contact MR measurement is shown in Fig. 4.6. The separation between the sample (M_1) and tip (M_2) is controlled by screws. A modified version uses a piezo system to control the separation. The magnetic field is applied by a pair of Helmholtz coils (1 and 2).

Measured results The PMRs of CrO_2 for a pressed powder and for a powder diluted with 75 wt% insulating Cr_2O_3 particles with similar shape were measured as a function of temperature. The results are shown in Fig. 4.7. At 5 K, PMR is approximately 50% for the diluted powder compact (Coey *et al.*, 2002). The value of MR for undiluted CrO_2 compact (black squares) is approximately 30%. The corresponding polarizations are 62% and 82%, respectively. However, it is only 0.1% at RT (Coey *et al.*, 2002).

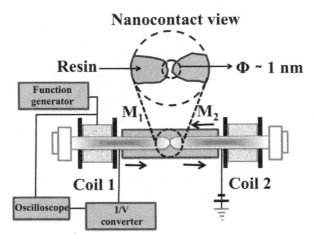

Fig. 4.6. Basic schematic setup of point-contact MR measurement (García *et al.*, 1999). M_1 and M_2 form the point contact. The battery connected to M_2 applies the bias voltage.

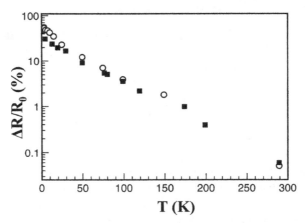

Fig. 4.7. PMR results for CrO_2 (black squares) and 25% powder compact of CrO_2 (open circles) (Coey *et al.*, 2002).

4.2.4.2. *Magnetization and magnetoresistance vs. applied external field*

The response of powder CrO_2 under a dc magnetic field up to 60 kG in terms of magnetization and MR was measured by Coey *et al.* (2002). The magnetization was measured by a SQUID magnetometer. These results at 5 K are shown in Fig. 4.8. The MR exhibits a butterfly pattern. The magnetization

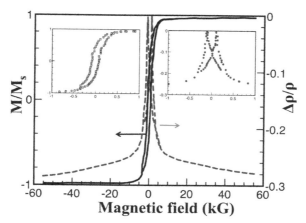

Fig. 4.8. Magnetization (solid curve) and MR (dashed curve) at 5 K as function of fields up to 60 kGauss (kG) for a CrO_2 powder compact (Coey *et al.*, 2002).

shows a hysteresis loop with coercivity 0.99 kG. At RT, it is 0.59 kG. The mass magnetization at saturation M_s is approximately 115 emu/g.

4.2.5. *Half-metallic properties*

While measuring the MR, it is possible to deduce the polarization P. For large P, it is possible that the sample is an HM. To confirm this, however, it is necessary to carry out Andreev reflection or other more direct measurement techniques.

4.2.5.1. *Point-contact Andreev reflection (PCAR)*

The basic principle of the Andreev reflection is discussed in Chapter 2. To study the reflection between an HM and a superconductor (SC), Soulen *et al.* (1998) developed the point-contact scheme (discussed in Section 4.2.4.1).

Anguelouch *et al.* (2001) used PCAR to analyze CrO_2. Lead (Pb) is used as SC and the sample CrO_2 is in film form. The advantages of using PCAR over the use of tunneling junctions and nanolithography are two-fold:

- The measurements are not fabrication-dependent.
- The measured results are independent of the characteristics of the interface between the sample and superconductor.

Tips and probes   The tips were obtained from wires 0.03 inch in diameter (Ji *et al.*, 2001), treated first by mechanical polishing and then electrochemical etching. Since the conductance as a function of voltage was to be determined, a conventional four-probe scheme (e.g., Fig. 2.20) was used.

Setup   The setup is shown in Fig. 2.33. The tip was controlled by a differential screw mechanism for microscopic movements toward the sample surface. To keep the measurements at low temperature, the whole setup was encased in a vacuum jacket. The jacket was immersed in a liquid helium bath. The temperature at the measurements was 1.85 K.

What is measured   The contact resistances or conductances dI/dV were measured by standard ac lock-in techniques. Anguelouch *et al.* (2001) analyzed their data and extracted the polarization by using the modified model (Strijkers *et al.*, 2001) of Blonder *et al.* (1982) — a model for understanding the conductance as a function of voltage measured from a point contact of nonmagnetic metal and superconductor (SC). Ji *et al.* (2001) reported their experimental results for $G(V)/G_n$ for the case $P_c = 96\%$ at $T = 1.85$ K (Fig. 4.9) which agree well with the calculated results using the same parameters. At very low temperature, CrO_2 shows close to 100% polarization.

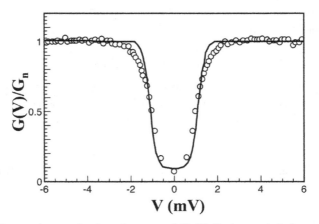

Fig. 4.9.   Measured normalized conductance of Pb/CrO_2 (open circles) as a function of V for $Z = 0$, $P_c = 0.96$, $\Delta = 1.12$ meV, $R = 4.0\,\Omega$, and $T = 1.85$ K (Ji *et al.*, 2001). The solid line is a fit to the modified model of Blonder, Tinkham and Klapwijk using the same parameters.

4.2.6. *Electronic properties*

In this section, we are concerned mainly with photoemission spectra. We shall start with the basic principles of conventional ARPES, then we shall discuss the spin-polarized photoemission, the measurements and setups, and finally the results for CrO_2.

4.2.6.1. *Spin-polarized angle-resolved photoemission spectroscopy (ARPES)*

Photoemission experiments measure the energy distribution curve (EDC): the intensity of the emitted electrons at a given angle and photon energy as a function of binding energy of the electrons with respect to E_F (see Section 2.4.3.1 for more details). A spin- and energy-resolved photoemission study of polycrystalline CrO_2 films was carried out by Kämper *et al.* in 1987. The light source was a He gas discharge lamp with photon energy $h\nu = 21.2\,eV$. The films were magnetized in the film plane using a coil with pulse field of about 500 Oe. The emitted electrons were detected normal to the film plane by means of a 180° spherical energy analyzer with energy and angle resolution of 100 meV and ±3°, respectively. In Fig. 4.10(a), the photoemission EDC of polycrystalline CrO_2 films measured at 300 K for three different sputter cleaning times (t_{sp}) are given (Kämper *et al.*, 1987). The sputter cleaning process changes the surface condition.

The most crucial information provided by the measurements is the appearance of the shoulder around 2 eV for $t_{sp} = 120\,s$. It persists up to $t_{sp} = 335\,s$. Figure 4.10(b) shows the corresponding spin-polarization spectrum for $t_{sp} = 120\,s$. Although the near 100% spin polarization was observed near 2 eV below E_F, they did not observe a large spin polarization near E_F.

Figure 4.11 shows the spin-polarized EDC at a particular **k**-point in the first BZ of a $CrO_2(100)$ surface measured by Dedkov (2004) for different t_{sp}'s. The experiments were carried out at RT. The important features of the results are summarized as follows:

- After $t_{sp} = 210\,s$, a shoulder developed at energy 2.3 eV below E_F for the total EDC. This feature agrees well with the earlier results of Kämper *et al.* (1987). A similar feature is exhibited in the ↑ spin channel. The ↓ spin states do not show any such structure. The polarization at E_F is 95%.
- After $t_{sp} = 750\,s$, the shoulder of the total EDC is shifted to 2.0 eV and its strength is increased.

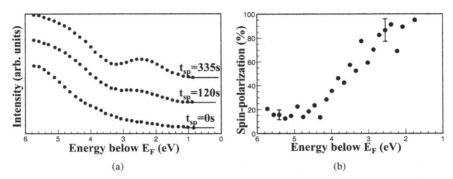

Fig. 4.10. (a) Photoemission EDC of polycrystalline CrO_2 films measured at 300 K for different times of sputter cleaning. (b) Spin-polarized spectrum of polycrystalline CrO_2 films at 300 K after $t_{sp} = 120\,s$ (Kämper *et al.*, 1987).

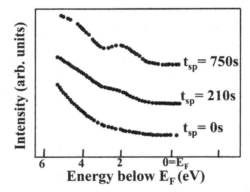

Fig. 4.11. Photoemission EDC of $CrO_2(100)$ surface for different sputter cleaning times (Dedkov, 2004). The energy scale is referenced to E_F.

- After $t_{sp} = 750\,s$ and additional annealing for 12 hours at $150\,°C$, the shoulder feature persists. The polarization at E_F is 85%.

In summary, the experimental results are interpreted as follows:

- The $-2.3\,eV$ shoulder is primarily from the 3d-states.
- The difference of the total EDC at $t_{sp} = 210\,s$ and $t_{sp} = 750\,s$ is due to the sputtering process which can either increase disorder near the surface of the sample or reduce d-p hybridization between the Cr and O atoms.
- $P \sim 85\%$ for $t_{sp} = 750\,s$ is due to the annealing process which recovered the quality of the surface and removed contaminants.

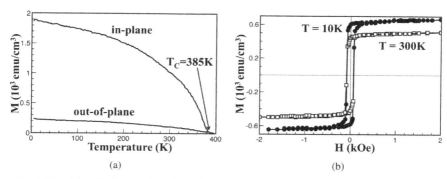

Fig. 4.12. (a) $M(T)$ as a function of T for thin-film form of CrO_2 on $Al_2O_3(0001)$ substrates (Kuneš *et al.*, 2002). (b) Hysteresis loops of $CrO_2(100)$ films epitaxially grown on a $TiO_2(100)$ substrate at $T = 10\,K$ (dots) and $300\,K$ (open squares) (Spinu *et al.*, 2000).

4.2.7. *Magnetic properties*

4.2.7.1. *Curie temperature T_C*

CrO_2 is an FM material. Most of the studies of the magnetic properties have used samples in thin-film form. Kuneš *et al.* (2002) measured temperature dependence of the in-plane and out-of-plane magnetizations in CrO_2 film grown by CVD on $Al_2O_3(0001)$ substrates. The easy axis is in the plane of the film. Both in-plane and out-of-plane magnetizations vanish at 386 K, as shown in Fig. 4.12(a), which led Kuneš *et al.* to conclude that T_C is 385 K. An earlier experiment (Kouvel and Rodbell, 1967) on compressed powdered stoichiometric CrO_2 reported a T_C of 386.5 K. These two results agree well with each other.

4.2.7.2. *Saturation magnetic moment*

The saturation magnetization for a thin film of CrO_2 at $T = 10\,K$ was found to be $600\,emu/cm^3$, and it is $450\,emu/cm^3$ at $T = 300\,K$ (Spinu *et al.*, 2000).

4.2.7.3. *Hysteresis loops*

The hysteresis loops of $CrO_2(100)$ films epitaxially grown on a $TiO_2(100)$ substrate were measured by a SQUID magnetometer with magnetic field applied along the easy axis of the films (Spinu *et al.*, 2000). The results are shown in Fig. 4.12(b).

As shown later in the band structure, because E_F does not intersect the d-bands, the magnetism cannot be of the Slater–Stoner–Wohlfarth type. The localized moments must be aligned by some kind of inter-atomic exchange interaction. One possibility is the superexchange through the oxygen atom.

4.2.8. *Theoretical studies of electronic and magnetic properties*

4.2.8.1. *Electronic properties*

Band structure and density of states   The band structure of CrO_2 has been calculated by many authors using first-principles methods based on DFT. Schwarz (1986) was the first to use the spin-polarized augmented spherical wave method of Williams *et al.* (1979). Later, LSDA/linear combination of Gaussian orbitals (LCGO) (Brener *et al.*, 2000), GGA/LAPW (Mazin *et al.*, 1999), and LSDA+U/LMTO (Korotin *et al.*, 1998) methods were applied. As an example, the ↑ (majority) and ↓ (minority) spin band structures calculated with the GGA exchange-correlation by Brener *et al.* (2000) are shown in Fig. 4.13.

All these calculations show that CrO_2 is an HM with a magnetic moment of $2.0\,\mu_B$/formula unit. Other properties, such as the insulating gap in the minority-spin channel and DOS of the majority-spin channel at E_F, are listed in Table 4.2. Shown in Fig. 4.14 is the DOS of the two channels based on the GGA exchange-correlation functional (Brener *et al.*, 2000).

Qualitatively, the three results given in Table 4.2 show a minimum in the DOS at E_F of the majority-spin channel. The presence of a

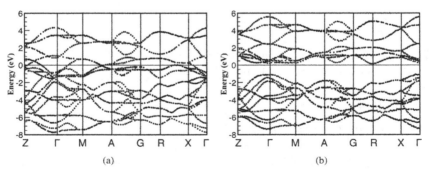

Fig. 4.13.   Band structure of CrO_2 (Brener *et al.*, 2000). (a) ↑ (majority) spin states; (b) ↓ (minority) spin states.

Table 4.2. Comparison of theoretical density of states.

DOS (states/eV-cell) at E_F Majority-spin states	Gap (eV) Minority-spin states	Reference
~0.5 (a minimum)	1.5	Schwarz (1986)
2.33 (close to a minimum)	1.34	Brener *et al.* (2000)
0.95 (a minimum)	1.3	Mazin *et al.* (1999)
~0.3 (a minimum)	1.5	Korotin *et al.* (1998)

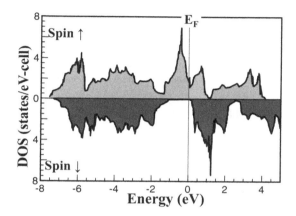

Fig. 4.14. Spin-projected densities of states of CrO_2 (Brener *et al.*, 2000).

minimum agrees reasonably well with the photoemission spectra. However, quantitatively, the amplitude of the DOS differs by nearly a factor of eight. The agreement of the gap values for minority-spin states is much better.

All the theoretical calculations agree that the d-states of the Cr split into triply (t_{2g}) and doubly (e_g) degenerate sets of states under the crystal field of the O atoms. The bonding properties of CrO_2 show both ionic character and d-p hybridization of more covalent character. Korotin *et al.* (1998) show that it is the e_g states originating from the Cr atom which contribute to the hybridization. The crystal field and hybridization combined with the exchange interaction gives rise to the half-metallic properties.

Fermi surface The Fermi surface of CrO_2 crystal is rather complicated. The results of Brener *et al.* (2000) are shown in Fig. 4.15 in different sections of the BZ. There is a roughly cubical electron surface around the Γ point with hole surfaces located between the electron surface and $\mathbf{k}_z$ zone boundaries. The points of hole surfaces closest to the Γ point touch

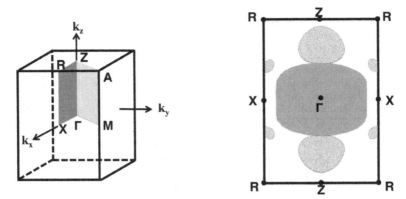

Fig. 4.15.   The BZ and Fermi surfaces in the section ΓZRX (Brener *et al.*, 2000). The gray region is the electron surface while the light gray regions are the hole surfaces.

the electron surface. These results differ from Mazin *et al.* (1999) in two aspects:

- The hole surface determined by Mazin *et al.* touches the A-Z plane and forms hammerheads.
- Mazin *et al.* also find needle-shaped surfaces.

Hubbard U   The Hubbard-U value obtained by Korotin *et al.* (1998) is 3.0 eV. The effect of U is to shift the unoccupied minority-spin states with respect to the LSDA results and to make the dip in the DOS at E_F of the majority-spin channel more pronounced. Given these effects, it is still controversial whether it is necessary and appropriate to determine the band structure of CrO_2 with LSDA+U.

4.2.8.2. *Magnetic properties*

Magnetic moment   The value of the magnetic moment was first explained by Schwarz using an ionic model — two oxygen atoms take a total of four electrons from the Cr atoms. This leaves two electrons at the Cr site. Hund's first rule requires the spin moments of the two electrons to be parallel. The spin angular momentum of an electron along a given axis is $\hbar/2$. With the g-factor of 2, the magnetic moment in the cell is then $2.0\,\mu_B$. Brener *et al.* (2000) object to the ionic model based on the calculations by Schwarz (1986) — the total number of electrons within the Wigner–Seitz sphere of the Cr atom being 4.2 rather than 2.0. The alternative explanation proposed by Brener *et al.* is that the ↑ and ↓ d-p hybridized states are fully occupied so that they do not contribute to the magnetic properties of the crystal. The remaining two electrons at the Cr site give the $2\,\mu_B$/unit cell.

Mechanism of ferromagnetism The band structure obtained by Korotin *et al.* (1998) shows that the d_{yz} and d_{zx} orbitals of the Cr atom hybridize with p-states of the O atoms. These hybridized states are mobile and can move into the region where the d_{xy} orbital is localized. Hund's first rule requires the three electron spins to be aligned, resulting in the FM coupling between the localized states. The mechanism is similar to Zener's double exchange mechanism (Zener, 1951). This will be discussed further in Section 4.5.2.

4.3. Fe₃O₄

Fe_3O_4 is sometimes called magnetite. It is a well-known magnetic material with a T_C of 858 K (van Dijken *et al.*, 2004). It has a more complicated crystalline structure than CrO_2. The Fe ions have two different valences: Fe^{2+} and Fe^{3+}. In addition, it exhibits a Verwey transition (Verwey, 1939): a sudden decrease of conductivity at $T_V = 120$ K, which is attributed to an order-disorder transition of Fe^{2+} and Fe^{3+} on the B sublattice and the accompanying distortion of the lattice.

4.3.1. *Structure*

The crystal structure above T_V has been determined by X-ray diffraction and its magnetic structure has been probed by neutron scattering (Shull *et al.*, 1951). It is the cubic spinel structure. The primitive lattice is fcc with lattice constant of 8.396 Å. Associated with each fcc lattice point, there are two smaller cubes, as shown in Fig. 4.16. The space group of the crystal is Fd3̄m. Each primitive fcc cell has the following configurations of atoms:

- Fe^{2+} sites: The unit cell contains two Fe^{2+} ions. One of them is at the origin of the unit cell — at the lower left corner (black polygon) — and the other one is at the center of one of the small cubes (open polygon). They are at the tetrahedral (A) sites.
- Fe^{3+} sites: The other Fe ions are 3+ and are located at four corners of the smaller cube (gray polygons) with a vacancy at its center. These Fe ions are at the octahedral (B) sites.

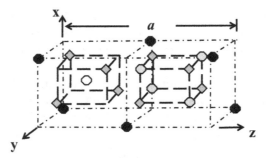

Fig. 4.16. Part of the spinel structure associated with one lattice point (lower left corner) of the conventional fcc structure. The tetrahedral (A) sites (black and open polygons), the octahedral (B) sites (gray polygons), and the oxygen atoms (diamonds) in Fe_3O_4.

Table 4.3. Atomic positions in reduced coordinates and effective charges of the Fe ions in Fe_3O_4.

Element	Atomic position	Effective charge (e)
Fe	(0.000, 0.000, 0.000)	2+
Fe	(0.250, 0.250, 0.250)	2+
Fe	(0.125, 0.125, 0.625)	3+
Fe	(0.375, 0.375, 0.625)	3+
Fe	(0.125, 0.375, 0.875)	3+
Fe	(0.375, 0.125, 0.875)	3+
O	(0.375, 0.125, 0.125)	
O	(0.125, 0.375, 0.125)	
O	(0.125, 0.125, 0.375)	
O	(0.375, 0.375, 0.375)	
O	(0.375, 0.125, 0.625)	
O	(0.125, 0.375, 0.625)	
O	(0.125, 0.125, 0.875)	
O	(0.375, 0.375, 0.875)	

- O sites: The O atoms occupy four corners of each small cube (diamonds).

In Table 4.3, we list the atom positions in reduced coordinates of the lattice constant a. The magnetic structure of Fe_3O_4 was first proposed by Nèel (1948). All Fe atoms at B sites are ferromagnetically aligned with each other and antiferromagnetically aligned with those at A sites. This structure was experimentally confirmed by Shull *et al.* (1951). Fe^{3+} ions on the B sublattice have magnetic moments of $5\,\mu_B$, while Fe^{2+} ions have magnetic moments of $4\,\mu_B$ and $5\,\mu_B$.

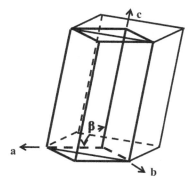

Fig. 4.17. The unit cell (outlined in thick lines) of Fe_3O_4 at temperature below T_V (Wright *et al.*, 2002).

Below T_V, Iizumi *et al.* (1982) determined the structure using the XRD method on a single crystal of synthetic Fe_3O_4. The unit cell determined at 10 K is monoclinic with three lattice constants along **a**-, **b**-, and **c**-directions being $\sqrt{2}a$, $\sqrt{2}a$, and $2a$, respectively, where a is 8.394 Å and is for the cubic phase above T_V. The corresponding space groups going from the cubic to monoclinic structure are $Fd\bar{3}m \rightarrow Cc$. In 2002, Wright *et al.* used high-resolution neutron and synchrotron X-ray powder-diffraction measurements to refine the structural information of Fe_3O_4 below T_V. The refined structure information of the monoclinic unit cell has the space group $P2/c$ instead of Cc with lattice parameters $a = 5.944$ Å, $b = 5.924$ Å, $c = 16.775$ Å, and angle $\beta = 90.2363°$. The unit cell is shown in Fig. 4.17. The positions of the atoms at 90 K were constrained by the operators of the orthorhombic space group $Pmca$ and are given in Table 4.4, with the origin at an inversion center located between two A sites.

4.3.2. *Growth*

Fe_3O_4 has been grown in single-crystal and thin-film form.

4.3.2.1. *Single-crystal growth*

McQueeney *et al.* (2006) have recently reported growing 5–10 g single-crystal samples. They used a radio frequency induction melting technique to grow the samples from the powdered Fe_3O_4. Initially, the powdered materials were in a crucible lined with a solid of the same composition to minimize the possible contamination of the melt. After slow cooling,

Table 4.4. Atomic positions of of Fe_3O_4 with Pmca symmetry at $T = 90$ K (Wright *et al.*, 2002).

Atom	x (1/a)	y (1/b)	z (1/c)
Fe (A_1)	0.25	0.0034	0.0637
Fe (A_2)	0.25	0.5061	0.1887
Fe (B_{1a})	0.00	0.5000	0.0000
Fe (B_{1b})	0.50	0.5000	0.0000
Fe (B_{2a})	0.00	0.0096	0.2500
Fe (B_{2b})	0.50	0.0096	0.2500
Fe (B_3)	0.25	0.2659	0.3801
Fe (B_4)	0.25	0.7520	0.3766
O (1)	0.25	0.2637	−0.0023
O (2)	0.25	0.7461	−0.0029
O (3)	0.25	0.2447	0.2542
O (4)	0.24	0.7738	0.2525
O (5a)	−0.0091	0.0095	0.1277
O (5b)	0.4909	0.0095	0.3723
O (6a)	−0.0081	0.5046	0.1246
O (6b)	0.4919	0.5046	0.3754

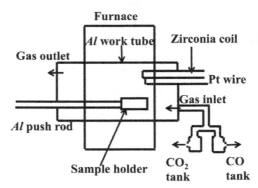

Fig. 4.18. A schematic diagram of the annealing chamber (Shepherd and Sandberg, 1984).

the crystals were extracted from the fractured boule. The single crystals were re-annealed under appropriate CO_2/CO atmospheres. The chamber for re-annealing was specially designed by Shepherd and Sandberg (1984). A schematic diagram of the annealing chamber is shown in Fig. 4.18. The chamber is made of alumina and is partly enclosed inside a furnace.

4.3.2.2. *Thin-film growth*

Many researchers have grown thin-film Fe_3O_4 using different methods. Here we comment on the PLD method and epitaxial growth.

Pulsed laser deposition Recently, Bollero *et al.* (2005) used PLD to grow 26–320 nm thick Fe_3O_4 films on $MgAl_2O_4(100)$ substrates. The lattice constant of the substrate is 8.083 Å. There is a -4% lattice mismatch between the substrate and samples. The films show antiphase domain boundaries. A Lambda-Physik 248 nm KrF excimer laser was used. Its repetition rate was 10 Hz and the fluence was 0.65 J/cm^2. The substrate was kept at 500 °C during the growth in a background of oxygen at pressure 5.0×10^{-6} mbar.

In 2003, Reisinger *et al.* reported the PLD growth of Fe_3O_4 on Si(111) substrates. Instead of using other oxides, such as MgO, as substrates, this experiment used Si as the substrate. They had in mind to integrate the magnetic properties of Fe_3O_4 into Si technologies.

The growths involved several stages, namely the preparation of Si substrates, PLD of buffer layers made of TiN and MgO to have a better lattice-constant match with the grown samples, and finally the deposition of Fe_3O_4. A schematic diagram is shown in Fig. 4.19.

It is advisable to have *in situ* characterization of the substrates and samples by RHEED during the thin-film growth. To grow the oxides, an

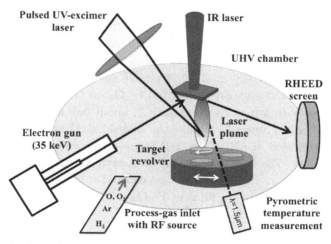

Fig. 4.19. A schematic of the apparatus for growing thin-film Fe_3O_4 (Reisinger *et al.*, 2003).

issue related to the *in situ* measurement needs to be addressed. The samples have to be grown in high oxygen partial pressure. The oxygen gases interact strongly with the high-energy electron beams. To prevent the scattering of the electron beam in the oxygen gas from causing misinterpretation of the results, Reisinger *et al.* (2003) built a RHEED system operated under pressure (10.0 Pa). In addition, it has been shown that it is advantageous to heat the substrate to high temperature in epitaxial growth of thin films. Ohashi *et al.* (1999) introduced a heating scheme using an infrared (IR) diode laser. With this scheme, the substrate temperature up to 1300 °C was provided by a laser beam with 95 μm wavelength and a maximum power of 100 W. The beam was focused on the backside of the substrate, in the region painted by silver. Temperature of the substrate was measured by a high-resolution pyrometer operating at 1.5 μm rather than a commonly used thermocouple.

The Si substrates were a commercial Si (001) single-crystal wafer. Aceton and isopropanol were used in an ultrasonic bath to pre-clean the substrates. A possible oxide layer can be formed during the cleaning procedures but can be removed by an *in situ* cleaning scheme using the IR laser.

A pulsed KrF excimer laser with wavelength 248 nm and energy density 2–5 J/cm^2 per shot was used to deposit TiN, MgO, and Fe_3O_4 on Si(001) surfaces. The repetition rate of the laser was between 2 and 10 Hz depending on the materials to be deposited.

After the growth of the buffer layers, Fe_3O_4 was epitaxially grown on top of MgO. Other than the repetition rate being changed to 2 Hz, the other conditions remained the same. The RHEED pattern indicated that the growth was layer by layer with smooth structure.

Laser ablation   Huang *et al.* (2002) used an ultraviolet (UV) laser to grow epitaxially 150 monolayers (ML) of Fe_3O_4 on MgO. MgO has a rocksalt structure with lattice constant 4.21 Å, about half that of the conventional cell of Fe_3O_4 above T_V. Before the growth, MgO was cleaved to MgO (100) and annealed at 650 °C for 1–2 hours under an oxygen pressure of 1.3×10^{-7} mbar. The base pressure in the growth chamber was maintained at 1.3×10^{-10} mbar. The Fe atoms were emitted from a water-cooled effusion cell. They formed thin films of Fe_3O_4 with the oxygen atoms under pressure in the order of 6.7×10^{-7} mbar.

Molecular beam epitaxy (MBE)   MBE growth of Fe_3O_4 films ranging from 3–100 nm in thickness was also reported by Eerenstein *et al.* (2002).

The background pressure of the growth chamber was 10^{-10} mbar. The oxygen pressure was 10^{-6} mbar during the growth.

van der Zaag *et al.* (2000) used the MBE method to grow Fe_3O_4/CoO bilayers on α-$Al_2O_3(0001)$ and $SrTiO_3(100)$ substrates. They used a Balzers UMS630 multichamber system. The deposit rate was between 0.02–0.05 nm/s at a substrate temperature of 523 K under oxygen pressure of 3×10^{-3} mbar.

A multilayer growth of Fe_3O_4 was reported by Cai *et al.* (1998) using a platinum (Pt) (111) substrate. The top of the Pt(111) surface was cleaned using bombardment of Ar^+ ions with energy 1.0 keV, then annealed to $T = 1350$ K under oxygen pressure of 10^{-7} mbar. A well-ordered surface was obtained. After the substrate surface was prepared, a submonolayer of Fe was grown on the surface by vapor deposition at RT and was heated for 2–3 min around 1000 K under oxygen pressure of 1.0×10^{-6} mbar. At this point, a stoichiometric bilayer of FeO(111) was obtained. Then, Fe deposition and oxidation cycles were repeated to form Fe_3O_4. The islands of Fe_3O_4 were grown on top of the FeO bilayer. The well-known Stranski–Krastanov (Stranski and Krastanow, 1939) growth mode — forming islands — dominates in the growth process. Finally, the islands of Fe_3O_4 coalesce into thick layers.

4.3.3. *Characterization*

Most layered forms of Fe_3O_4 have been characterized by RHEED, transmission electron microscopy (TEM), and STM methods. Since we have discussed the growth of Fe_3O_4 by PLD, we shall discuss the characterizations based on the same work.

The characterizations discussed by Reisinger *et al.* (2003) involve several stages.

4.3.3.1. *Si substrate*

The Si substrate was heated to 600 °C, then treated at $T_{sub} = 900$ °C and 1150 °C to remove amorphous Si oxides on Si(001), and eventually cooled back down to 600 °C for the growth of the buffer layers. RHEED patterns were taken at the [110] azimuth. The sequence of the patterns is shown in Fig. 4.20.

Figure 4.20(a) shows the RHEED pattern before the heat treatment. It shows a homogeneous background and a few spots. The background is

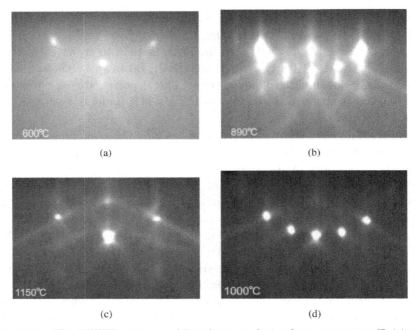

Fig. 4.20. The RHEED patterns of Si substrates during heat treatments (Reisinger *et al.*, 2003). (a) $T_{sub} = 600\,°C$; (b) after heating to $890\,°C$; (c) $1150\,°C$ (the maximum temperature); and (d) $1000\,°C$.

due to the presence of amorphous Si oxides. The spots are the constructive interference pattern of Si(001) surfaces. As the temperature increases to about $900\,°C$, it exhibits repeating long and short lines (Fig. 4.20(b)), indicating a 2×1 reconstruction, with the lines somewhat streaky. The homogeneous background is no longer visible. As the temperature increases to $1150\,°C$, the pattern characterizing the 2×1 reconstruction disappears. This suggests that the dimmer bonds forming the 2×1 reconstruction have dissolved. When the Si substrates are cooled down to $1000\,°C$, sharp Laue spots are clearly visible, indicating that the oxides have been removed. The five spots indicate that a clean 2×1 surface has been obtained.

4.3.3.2. *TiN buffer layer*

The RHEED patterns of the TiN films depend on the number of laser pulses at constant deposition temperature. After 70 pulses at $T_{sub} = 600\,°C$, the film thickness is $0.4\,nm$. The pattern shown in the left panel of Fig. 4.21 exhibits varieties of spots due to the growth of superstructures. When the

(a)   (b)

Fig. 4.21. The RHEED pattern of TiN on a Si(001) substrate at $T_{sub} = 600\,^{\circ}C$ (Reisinger *et al.*, 2003). (a) 70 pulses and (b) 1000 pulses.

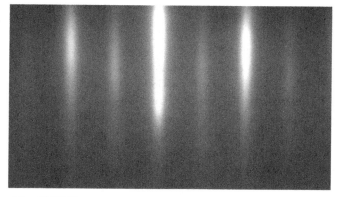

Fig. 4.22. The RHEED pattern of Fe_3O_4 on a TiN buffer layer after 5000 pulses (Reisinger *et al.*, 2003).

film is thicker than 0.8 nm, the pattern changes to a few strong elongated spots, indicating the growth of islands (right panel).

4.3.3.3. *Fe₃O₄ film*

Finally, the growth of Fe_3O_4 on the buffer layer was characterized by RHEED after 5000 pulses. The thickness of the magnetite was 40 nm. As shown in Fig. 4.22, a well-defined pattern with stripes is observed. The spacing between the stripes is half of that shown in the case of TiN. This is consistent with the fact that the lattice constant of Fe_3O_4 is twice that of TiN.

4.3.4. *Physical properties*

Fe_3O_4 has two attractive features:

- It has a ferrimagnetic transition temperature of 860 K (Smit and Wijn, 1959).
- Crystallites of the magnetite can be grown. It was predicted to show half-metallic property by Yanase and Siratori (1984) and Zhang and Satpathy (1991).

In this section, we discuss first the most relevant property to this monograph — the half-metallicity of Fe_3O_4. What is the experimental evidence? Then, we shall look into some surface issues. Finally, we shall discuss theoretical studies of the electronic properties of Fe_3O_4 and the mechanism for its magnetic properties.

4.3.4.1. *Half-metallic properties*

Evidence for half-metallic FM properties of Fe_3O_4 is furnished by an experiment of the spin-polarized photoemission measurements (Dedkov *et al.*, 2002). The experiment was carried out using the photon energy of 21.2 eV. The samples are epitaxially grown $Fe_3O_4(111)$ films. The temperature at the measurements was RT and the pressure in the chamber was 1.0×10^{-10} mbar. The energy resolution was 100 meV and the angular resolution was $\pm 3°$. An external pulse of magnetic field of 500 Oe is applied along the easy axis, in-plane $\langle 1\bar{1}0 \rangle$, of the films.

The spectra were obtained in normal emission by a 180° hemispherical energy analyzer connected to a 100 KeV Mott detector. Two sets of data are collected. One is the EDC $N(E)$ and the other is the polarization $P(E)$. With the spin-polarized case, the EDC is proportional to $I(\uparrow) + I(\downarrow)$. The polarization is proportional to $I(\uparrow) - I(\downarrow)$. In general, the structures in the EDC are not prominently exhibited. Therefore, sometimes additional measurements of the second-order derivative $d^2N(E)/dE^2$ need to be carried out. The polarization P is $-(80 \pm 5)\%$ for 300 layers of oxidized Fe after annealing at 250 °C (Fig. 4.23). This led Dedkov *et al.* (2002) to conclude that Fe_3O_4 is an HM.

4.3.4.2. *Magnetic properties*

The magnetotransport properties of Fe_3O_4 have been probed by several researchers. We separate the discussions into MR, junction MR (JMR),

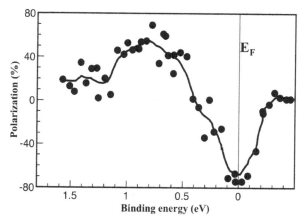

Fig. 4.23. Spin polarization as a function of binding energy for a 300-layer oxidized Fe after annealing at 250 °C (Dedkov *et al.*, 2002). The *P* value is $-(80\pm5)\%$. The thin solid line through the data points was fitted by a three-point averaging fast Fourier transformation procedure.

more recent MR measurements using nanocontacts and spin-valve structures, and, finally, neutron measurements, to understand the underlying mechanism of the ferromagnetism.

Magnetoresistance For polycrystalline thin-film samples of Fe_3O_4, Feng *et al.* reported, as early as 1975, the observation of negative MR for temperatures between 105 and 250 K. At 130 K and 23 kG, the MR reaches 17%. The sign of the MR has physical significance. If it is positive, the majority-spin polarization at E_F of the sample is parallel to the magnetization of the bulk or some reference systems, such as a trilayer. If it is negative, then the minority-spin polarization at E_F is parallel to the magnetization of the bulk. Gridin *et al.* (1996) examined single crystals of Fe_3O_4 and obtained a sharp peak near T_V with a maximum of 17% under a magnetic field of 77 kG. About the same time, Gong *et al.* (1997) measured the MR of Fe_3O_4 samples grown epitaxially on MgO and observed a similar peak around T_V (~119 K). In addition, they found that the MR increases monotonically as the temperature decreases below 105 K. For a film of 6600 Å thick, the MR is 32% at 60 K and field of 40 kG. The results are shown in Fig. 4.24. The dashed lines are the fit using Mott's formula, Eq. (4.8), based on the mechanism of variable range hopping.

$$\rho = Ae^{(\frac{B}{T})^{\frac{1}{4}}}, \tag{4.8}$$

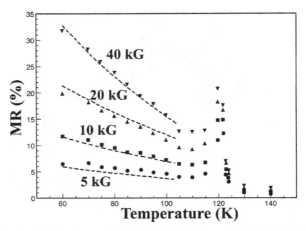

Fig. 4.24. Measured MR at magnetic field of 40 kG (down triangles), 20 kG (triangles), 10 kG (squares), and 5 kG (dots) in the temperature range 60–140 K for a Fe_3O_4 film. Fits to Eq. (4.8) are indicated by dashed lines (Gong *et al.*, 1997).

where A and B are parameters. B can be interpreted as the activation energy and is 1.61×10^8 K by fitting the zero-field case. Based on these fits, the authors attributed the origin of the MR increasing with decreasing temperature to a magnetic-field dependence of the activation energy for hopping conduction below T_V.

Besides the effects of the Verwey transition on the MR, Ziese and Blythe (2000) found also anisotropic MR (AMR)[2] in thin-film forms of Fe_3O_4 grown on MgO. The thicknesses of the films were 200, 50, and 15 nm, respectively, with 10% uncertainty. The current was along the [100] axis and the magnetic fields were applied along the [100] and [010] directions. For the thickest films, additional measurements were made with the current flowing in the [110] direction and magnetic fields oriented along the [110] and [1$\bar{1}$0] axes. The four-point geometry with silver-paste contacts was used to measure the resistivity. The MR was determined with respect to the resistivity at the coercive field at which there is no magnetization in the sample. A typical result of the MR showing the anisotropy is given in Fig. 4.25. It is clear that the parallel and perpendicular MRs are different. However, clear also is the anisotropy of each. Ziese and Blythe attributed the anisotropy of the MR to the anisotropy of the samples.

---

[2]The background of AMR is discussed in Appendix A.

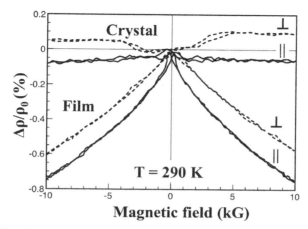

Fig. 4.25. The MR ratio $\Delta\rho/\rho_0$ of the Fe_3O_4 single crystal and 200 nm thick film. The solid lines are measurements in longitudinal geometry and the dashed lines in transverse geometry (Ziese and Blythe, 2000).

Junction (tunneling) magnetoresistance The MR can also be measured by either the magnetic junction (Hu and Suzuki, 2002) or the nanocontact method (Versluijs and Coey, 2001). Occasionally, the results obtained from the former method are called the junction magnetoresistance (JMR). Hu and Suzuki measured the JMR of a trilayer junction made of $Fe_3O_4/CoCr_2O_4/La_{0.7}Sr_{0.3}MnO_3$ (LSMO). $CoCr_2O_4$ serves as a weak paramagnetic barrier (for $T > 95\,K$) and has 0.8% lattice mismatch with Fe_3O_4. The trilayers can be grown on (110) $SrTiO_3$ by PLD. LSMO grown on 1900 Å $SrTiO_3$ exhibits FM half-metallic properties with a T_C of 360 K (Park *et al.*, 1998a). Between 0.4 eV (below E_F) and E_F, the spin-polarized photoemission spectra measured at 40 K show ~100% polarization. Therefore, it is used as a spin analyzer for Fe_3O_4.

The JMR measurements were carried out with the current flowing in the direction perpendicular to the plane and the external magnetic field parallel to the plane of the film, along the easy [001] axis. To ensure that there is no exchange coupling between the magnetized layers, the magnetization as a function of magnetic field (H) was measured. The result of an unpatterned trilayer at 80 K is shown in Fig. 4.26 and indicates that coercive fields of the magnetic layers, LSMO and Fe_3O_4, are clearly separated at 280 Oe and 1.0 kOe, respectively. Figure 4.27 shows JMR as a function of H at 80 K for a 20 μm × 20 μm cross section. Using the resistance measured at 4.0 kOe

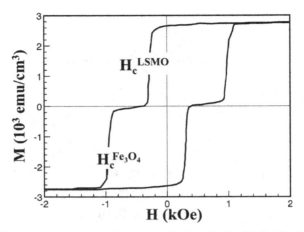

Fig. 4.26. The magnetization as a function of H of an $Fe_3O_4/CoCr_2O_4/LSMO$ trilayer (Hu and Suzuki, 2002).

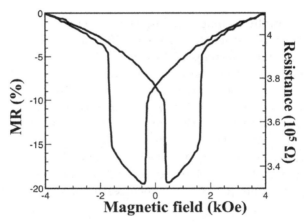

Fig. 4.27. The JMR defined in Eq. (4.9) as a function of H for an $Fe_3O_4/CoCr_2O_4/$ LSMO trilayer ($T = 80$ K) (Hu and Suzuki, 2002).

(R_{4kOe}) as a reference, JMR is defined as

$$\text{JMR} = \frac{R_{4kOe} - R_H}{R_{4kOe}}. \tag{4.9}$$

The results show negative JMR with the maximum JMR at -20%.

Magnetoresistance determined by nanocontacts Transport measurements using nanocontacts were developed after the STM method. One notable feature is the possible observation of quantized conductance. Versluijs and Coey (2001) applied this method to Fe_3O_4. They used two magnetite crystals with volumes less than $0.1 \, mm^3$. These crystals were brought together into mechanical contact and were then pulled apart by a piezo stack. The experiments were done at ambient conditions. A set of Helmholtz coils was used to apply a magnetic field of $70 \, G$ in the parallel and perpendicular directions of the current. There was a bias voltage up to $\pm 0.3 \, V$ applied between the two crystals. An I-V converter with a low-noise amplifier (OP37) was used to measure the current. After the amplification, the current was fed into a PC ADC (analog-to-digital converter) card for analysis.

The results do not show quantized conductance. As the crystals were slowly pulled apart, the resistance was continuously increased. The I-V curves at two magnetic fields, $B = 0$ and $B = 70 \, G$, are shown in Fig. 4.28 with the curves fitted with $I = GV + cV^3$, where G and c are fitting parameters. The MR was defined similarly to the JMR (Eq. (4.9)) with reference R at $B = 0$. The MR as a function of conductance is shown in Fig. 4.29. The maximum MR is 85% for resistive contacts and the MR decreases as the conductance increases. The authors ruled out that the behavior of the MR is caused by magnetostrictive effects. They attributed it to the formation of a domain wall at the nanocontact based on the nonlinear results of the I-V measurements.

Magnetoresistance of spin valves involving Fe_3O_4 Spin valves are made of two magnetic layers separated by a normal metal. The negative MR shown by other experiments suggests that Fe_3O_4 in spin-valve configurations could yield GMR. This could be a significant technological advance. van Dijken *et al.* (2004) fabricated two spin valves made of $Fe_3O_4/Au/Fe_3O_4$ and $Fe_3O_4/Au/Fe/Au$, respectively, by dc-magnetron sputtering on MgO(001) substrates under a base pressure of $10^{-7} \, mbar$. The samples had $10 \times 10 \, mm^2$ areas. The current was measured in a van der Pauw configuration with contacts at the four corners of the surface of the sample. The magnetic field was applied in-plane with parallel and perpendicular directions of the current. The magnetization curves were determined using a SQUID magnetometer.

In Fig. 4.30, the longitudinal and transverse MR of 30 nm–Fe_3 O_4/5 nm–Au/10 nm–Fe_3O_4 on MgO(001) spin valve, the first spin valve,

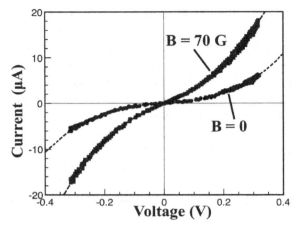

Fig. 4.28.   The *I-V* curves of Fe_3O_4 nanocontacts. The dashed lines are the fitted results (Versluijs and Coey, 2001).

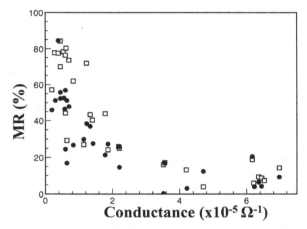

Fig. 4.29.   MR as a function of conductance at bias voltage of 0.3 V (filled circles) and zero bias (open squares) for Fe_3O_4 nanocontacts (Versluijs and Coey, 2001).

as a function of applied magnetic field are shown in Fig. 4.30(b). For comparison, data for a 50 nm Fe_3O_4 film on MgO(001) are shown in Fig. 4.30(a). Figure 4.30(c) is for 30 nm–Fe_3O_4/5 nm–Au/10 nm–Fe/2 nm–Au spin valve on MgO(001), the second spin valve. The measurements were carried out at 300 K. The GMR and AMR are much reduced in both spin-valve cases, indicating current passing through the Au region. In fact, no

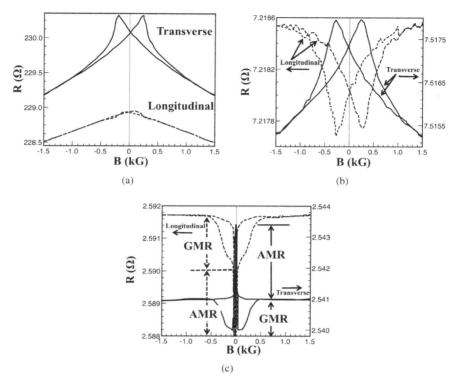

Fig. 4.30. MR as a function of magnetic field for Fe_3O_4 grown on MgO(001) (a); and for the two spin valves (b) and (c) (van Dijken *et al.*, 2004).

GMR is observed in the first spin valve due to the reversal of the magnetization in each Fe_3O_4 layer (van Dijken *et al.*, 2004). The AMR of the second spin valve is more pronounced. Because all the contacts were made on the surface of the top Fe_3O_4 layer, it is reasonable to conclude that the AMR originates in this Fe_3O_4 region and possibly from the structural and stoichiometric differences in the regions between Fe_3O_4 and Au and between Fe_3O_4 and MgO substrate. The second spin valve shows quite different MR behavior as compared to the other two cases, showing an isotropic GMR at low external magnetic field. The AMR of the second spin valve is shown by the decrease of the longitudinal MR but increase of the transverse MR as the magnetic field is cycled between -1 to $1\,$kG. van Dijken *et al.* suggested that this is due to the reversal of magnetization in the Fe layer. They also attributed the isotropic GMR in the second spin valve to the independent magnetizations in the Fe_3O_4 and Fe layers.

4.3.4.3. *Charge ordering*

The Verwey transition has a charge-ordered ground state at $T_V \sim 120\,\mathrm{K}$. The crystal exhibits an ordered mixed valence in the Fe ions at the B sites. According to Nazarenko *et al.* (2006) there are six non-equivalent iron atoms. Two are at the tetrahedral (A) sites and four are at the octahedral (B) sites. The charge ordering happens at the B sites. Two of them, Fe (B_1) and Fe (B_2), are at the centers of oxygen octahedra. Fe (B_3) and Fe (B_4) are off-center.

Resonance X-ray diffraction was used because the technique yields both the site selective diffraction and the local absorption spectroscopy regarding atomic species. Reflections in the range of tens of electron volts can be recorded around the absorption edge of an element or elements where there is evidence showing strong energy and angular dependencies (Nazarenko *et al.*, 2006). Physically, it involves virtual photon absorption-emission by an electron initially occupying a core state and making a resonant transition to some intermediate state near E_F. This technique is particularly effective for probing charge, orbital or spin orderings under distortions. For example, different resonance frequencies can be correlated to the charge difference at atoms suffering dissimilar distortions. By measuring the shift of the absorption edges, the valences of the atoms were determined to be 5.38, 5.62, 5.40, 5.60 e for Fe (B_1) to Fe (B_4), respectively. On the other hand, Wright *et al.* (2001) deduced from their neutron and X-ray measurements the valences of the four Fe ions to be 5.6, 5.4, 5.4, and 5.6 e, respectively. Therefore, there are some discrepancies for Fe (B_1) and Fe (B_2) between these two measurements.

4.3.4.4. *Surface properties*

Because of the intense interest in magnetic thin films and multilayers, Kim *et al.* (2000b) investigated the surface properties of Fe_3O_4 (111) by growing the sample on Fe(110) *in situ*. They first used XAS, MCD, and LEED — allowing them to determine the ionic state and element-specific magnetic information. This combination of experimental methods identified unambiguously the magnetic overlayer to be ferrimagnetic Fe_3O_4 when the layer thickness exceeds 600 layers. Furthermore, they found that the polarization in Fe_3O_4 layers is in the opposite direction of that in the Fe substrate.

These authors also studied bilayer forms of $Fe_3O_4(111)/Fe(110)$ by LEED, spin-polarized photoemission spectroscopy (SPPS), and magnetic

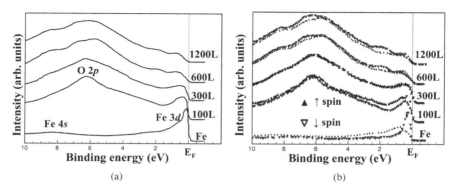

Fig. 4.31. (a) Intensity of photoemission spectra as a function of binding energy for different layer thicknesses (Kim *et al.*, 2000a). (b) The corresponding spin-resolved curves. The spectra for the pure Fe are shown at the bottom of both panels. The binding energy is measured with respect to E_F.

linear dichroism (MLD) (Kim *et al.*, 2000a). The photon energy was 120 eV. The spin- and angle-resolved photoelectrons were analyzed using a commercial VSW 50-mm spherical analyzer coupled with a low-energy diffuse scattering spin polarimeter. The resolution of the electronic energy was 0.1 eV and the angular resolution was around 2°. The measurement chamber was maintained at the base pressure of 1.3×10^{-11} mbar. The sample was annealed at 250 °C and was magnetized along the in-plane [001] direction, which is the easy axis of the thick Fe film.

Figure 4.31 shows the experimental spin-polarized photoemission spectra. In Fig. 4.31(a), the integrated spectra of the ↑ and ↓ spins are given as a function of binding energy at different thickness. Figure 4.31(b) shows the spin-resolved spectra. The pure Fe spectra are shown at the bottom of the figures. The binding energy is measured with respect to E_F.

For the pure Fe case, the width of d-states is approximately 3.0 eV. The weak peak at about 8.0 eV below E_F is attributed to s-states of Fe. As the Fe_3O_4 grows to 100 layers, a strong peak at 6.0 eV appears. It originates from p-states of the O atoms. When the thickness increases to 600 layers, several fine structures are present. By examining the spin-resolved spectra in Fig. 4.31(b), it is clear that the fine structures are due to magnetic ordering. It is also clear that the samples are not magnetic when the films are less than 300 layers in thickness.

The MLD spectra reveal the magnetic characteristics of the samples. In particular, the difference between spectra with opposite magnetizations can

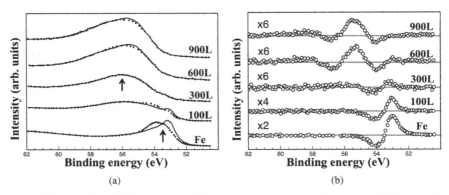

Fig. 4.32. (a) The MLD spectra and (b) the corresponding differences of MLD signals for four films of different thickness as a function of binding energy (Kim *et al.*, 2000a).

provide interesting information. In Fig. 4.32, the MLD spectra and differences between magnetizations of four films of different thickness are plotted with respect to the binding energy. In the MLD spectra, the pure Fe sample shows a narrow peak at around 53.5 eV. As the thickness increases, the peak becomes broader because in Fe_3O_4 the Fe ions have two different charges, Fe^{2+} and Fe^{3+}. The difference spectra indicate antiparallel coupling between the oxide layers and the Fe substrate. This can be seen in Fig. 4.32(b). For the Fe, the difference spectrum shows a positive slope at 53.0 eV. The 900-layer case (top curve) shows negative slope at around 54.0 eV. The authors confirmed this by growing more Fe on top of Fe_3O_4. As the Fe film thickness increases (1.2 ML), the onset of MLD changes its sign: i.e., the Fe metal and thick oxide overlayer have opposite magnetization. Kim *et al.* (2000a) attributed this to an antiparallel magnetic coupling in the $Fe_3O_4(111)/Fe(110)$ bilayer. This essential feature can be seen in a schematic diagram (Fig. 4.33). As compared to SPPS for the bulk-terminated surface of Fe_3O_4, there are two important features:

- The intensity of the photoemission spectra at E_F is higher in the films grown on Fe(110).
- The film samples show lower spin polarization (16%) than the bulk-terminated surfaces (40–60%).

There is not yet a convincing explanation of the intensity difference. One possible reason for the smaller polarization in the film samples is the lack of perfect stoichiometry.

Fe (110) Fe_3O_4 (111)   Fe (110)

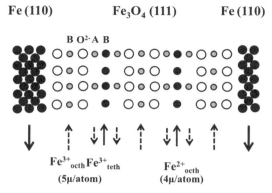

Fig. 4.33. Schematic diagram showing the antiparallel magnetization of the $Fe_3O_4(111)/Fe(110)$ structure. The arrows indicate the local direction of magnetization in the Fe planes. In Fe_3O_4 the magnetization is supported by the Fe^{2+}_{octh} ions (solid arrows), while the Fe^{3+} magnetic moments cancel (dashed arrows) (Kim *et al.*, 2000a).

4.3.5. *One-electron theory*

Zhang and Satpathy reported the first calculations of the spin-polarized band structure of Fe_3O_4 in 1991. They used the linear muffin-tin orbital method within the atomic sphere approximation (LMTO-ASA) (Andersen, 1975) and von Barth and Hedin exchange-correlation functional (von Barth and Hedin, 1972). The LMTO-ASA method is extensively discussed by Skriver (1983).

The Fe_3O_4 structure is shown in Fig. 4.16. A lattice constant of 8.397 Å and 18 empty spheres were used. In Fig. 4.16, the Fe atoms are located at the A and B sites. In terms of crystallographic terminology (Shepherd and Sandberg, 1984), they are at $8a$ and $16d$ sites, respectively. The frozen core approximation was applied to 3p-states of Fe and 1s-states of O. We summarize the sites and muffin-tin sphere radii used in Table 4.5. Muffin-tin orbitals of s, p, d on the Fe and O atoms and s and p type on the empty spheres were used as basis functions. The total charge density was constructed using a total of 56 **k**-points in the irreducible part of BZ.

There are three important contributions given by the LMTO-ASA calculations:

- Fe_3O_4 is predicted as an HM by the spin-polarized band structure and DOS.
- The relation between the non-spin-polarized band structure and the Stoner model.

Table 4.5.   Atomic site index and muffin-tin sphere radii of
Fe_3O_4. The site index is in crystallographic terminologies.

Atom	Site index	Muffin-tin sphere radius(Å)
Fe at A site	8a	0.95
Fe at B site	16d	1.15
O	32e ($x = 0.379$)	1.20
Empty sphere 1	16c	1.12
Empty sphere 2	8b	0.80
Empty sphere 3	48f ($x = 0.25$)	0.84

- The Verwey transition was studied by a three-band model Hamiltonian
  with all the parameters determined from first-principles calculations.

We shall discuss further the first two contributions, since we are interested
mainly in the half-metallic and magnetic properties.

4.3.5.1. *Spin-polarized band structure and DOS*

The spin-polarized band structure and corresponding DOS (Fig. 4.34) in
Fe_3O_4 were calculated by Zhang and Satpathy (1991). Key features in these
results are:

- The half-metallic behavior is characterized by the intersection of E_F with
  states in the minority-spin channel only. The states at E_F are identified
  as the t_{2g} states of the Fe atoms at B sites. In fact, the trigonal crystal
  field at the B site splits the t_{2g} states into doubly and singly degenerate states. e_g states originating from $d_{x^2-y^2}$ and d_{z^2} ($d_{3z^2-r^2}$) states
  form those states 2.0 eV above E_F. The Fe atoms at A sites do not contribute significantly to those states near E_F. They are occupied states
  with binding energy of about 2.0 eV with respect to E_F. The gap in the
  majority-spin states is approximately 1.8 eV.
- The p-bands associated with the O atoms are separated by a gap of
  about 1.0 eV from d-states of Fe atoms at A sites in the metallic channel,
  while the corresponding p- and d-bands in the majority-spin channel are
  separated by a small gap.
- The calculated magnetic moment M is $4.0\,\mu_B/Fe_3O_4$. For each Fe atom,
  the calculated M is $3.5\,\mu_B$. This latter value is considered an estimate
  because it depends on the chosen MT radii. The moments on the sublattices A and B are aligned antiferromagnetically.

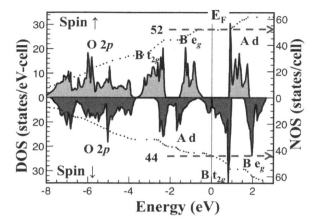

Fig. 4.34. The spin-projected densities of states (DOS) and numbers of states for magnetites (NOS, dot lines) of the majority- and minority-spin channels in Fe_3O_4 (Zhang and Satpathy, 1991). The net magnetic moment is $52 - 44 = 8\mu_B$ per unit cell which consists of two formula units.

Comparisons to experiment The integer M and finite DOS at E_F in the $\downarrow$ spin states show that Fe_3O_4 is an HM. The metallic behavior of this HM arises from the minority-spin channel. This is consistent with the negative MR observed in the experiments.

Polarized neutron scattering experiments have determined M at the A site to be $3.82\,\mu_B$. The calculated M at the A site discussed above agrees reasonably well with the experiments.

One point not discussed by Zhang and Satpathy is the extent of d-p hybridization between the Fe atoms at A and B sites with the O atoms. Based on the DOS shown in Fig. 4.34, Fe atoms at B sites hybridize with O atoms more strongly than those at A sites.

4.3.5.2. *Non-spin-polarized band structure and Stoner model*

The Stoner model (Stoner, 1939) explains successfully and intuitively the ferromagnetism in Fe. The essence of the model characterizing the magnetism in an FM metal is a consequence of the Pauli principle. This results in two competing energies. The first is the increase in kinetic energy as electrons are forced to occupy higher energy states and the second is the reduction of the Coulomb repulsion. To see how the non-spin-polarized band structure can be used in the Stoner theory,[3] we let $n = n_\uparrow + n_\downarrow$ be the total number of d-electrons/atom and $m = n_\uparrow - n_\downarrow$ be the magnetization/atom in the crystal. The net energy of the two competing energies is given by

$$E(m) = \frac{1}{2} \int_0^m dm' \left(\frac{m'}{N(n,m')} - I_{n_\uparrow n_\downarrow} \frac{m'^2}{4} \right), \qquad (4.10)$$

where $I_{n_\uparrow n_\downarrow}$ is the Stoner parameter. Physically, it is a measure of the strength of the intra-atomic Coulomb interaction between electrons contributing to the magnetization and is part of the Hubbard U (Kim, 1999). But in the present context, it is defined under the condition of a finite magnetization. Here,

$$N(n,m) = \frac{m}{\varepsilon_\uparrow - \varepsilon_\downarrow} \qquad (4.11)$$

is the DOS/atom-spin averaged over the spin-flip gap $\varepsilon_\uparrow - \varepsilon_\downarrow$. Where $\varepsilon_\uparrow$ and $\varepsilon_\downarrow$ are two energies around E_F for the $\uparrow$ spin occupancy $n_\uparrow = \frac{1}{2}(n+m)$ and $\downarrow$ spin occupancy $n_\downarrow = \frac{1}{2}(n-m)$ in units of states/atom-spin. The solution with net magnetization m_0 can be obtained by minimizing the energy $E(m)$ with respect to m. The condition for an FM state is

$$I_{n_\uparrow n_\downarrow} N(n,m_0) = 1. \qquad (4.12)$$

This is the Stoner criterion. In addition, the stable FM state requires the second-order derivative of $E(m)$ to be negative. Otherwise, the FM state is metastable. Equation (4.11) can also be derived from considering the chemical potentials for the $\uparrow$ and $\downarrow$ spin subsystems. The chemical potential for either spin channel is defined by the energy required to add or remove an electron from the corresponding spin band. These two chemical potentials are expressed as

$$\mu_\sigma = \varepsilon_\sigma + I_{n_\uparrow n_\downarrow} n_{-\sigma}. \qquad (4.13)$$

These chemical potentials suggest that the exchange interaction favors the addition of a spin state to the majority-spin band. Equation (4.11) is obtained by equating the two chemical potentials.

The Stoner model can be generalized for Fe_3O_4, where there are two sublattices (A and B) for the Fe atoms. With the two sublattices, it is possible that the number of d-electrons can change due to the magnetization.

---

[3]Marcus and Moruzzi (1988) have generalized the Stoner model to reproduce spin-polarized band structure calculations.

Thus, the two chemical potentials at site A can be expressed as:

$$\mu_{\uparrow/\downarrow}^A = \varepsilon_{\uparrow/\downarrow}(n_0^A, m^A) + (n^A - n_0^A)U^A + \frac{I_{n_\uparrow n_\downarrow}^A}{2}(n^A \mp m^A), \qquad (4.14)$$

where n_0^A (n^A) is the total number of d-electrons at site A before (after) the system exhibits magnetization. U^A is the on-site Coulomb repulsion energy at site A. A similar set of equations can be written for site B. Since the two sites are not distinguished in Eq. (4.12), the present theory cannot predict the relative orientation of the magnetizations at the two sites. The Stoner conditions for the A and B sites are:

$$I_{n_\uparrow n_\downarrow}^A N^A(n^A, m^A) = 1$$
$$I_{n_\uparrow n_\downarrow}^B N^B(n^B, m^B) = 1. \qquad (4.15)$$

By equating the two chemical potentials, $\mu_\uparrow$ and $\mu_\downarrow$, at each site, one has:

$$\varepsilon_\uparrow^A + \varepsilon_\downarrow^A + 2(n^A - n_0^A)U^A + I_{n_\uparrow n_\downarrow}^A n^A$$
$$= \varepsilon_\uparrow^B + \varepsilon_\downarrow^B + 2(n^B - n_0^B)U^B + I_{n_\uparrow n_\downarrow}^B n^B. \qquad (4.16)$$

There is also the condition of conservation of the total electron number with and without the magnetization.

$$n^A + 2n^B = n_0^A + 2n_0^B. \qquad (4.17)$$

The factor of 2 is due to the fact that there are twice as many Fe atoms at B sites. The ns and ms are obtained by solving Eqs. (4.13)–(4.15) self-consistently. Instead of carrying out self-consistent calculations, one can estimate the magnetizations at the A and B sites. In Eq. (4.14), the quantity U is the largest among all the terms. To satisfy Eq. (4.17), $n^A = n_0^A$ and $n^B = n_0^B$ are obtained. The Stoner parameters can be calculated by using the scheme of Poulsen *et al.* (1976). They turn out to be between 60–80 meV. The magnetizations at the two sites m^A and m^B are determined by using the calculated $N(n, m)$ for the two sites from the non-spin-polarized band structure. m^A and m^B are approximately 3.5 μ_B.

4.3.5.3. *Local spin density approximation with and without U*

Anisimov *et al.* (1996, 1997) used the LMTO method to calculate the band structure of Fe$_3$O$_4$ with LSDA and LSDA+U, respectively. More recently, Jeng *et al.* (2004) carried out similar calculations using the VASP implementation of the PAW method.

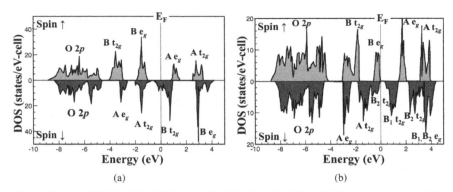

Fig. 4.35. (a) PDOS with LSDA, and (b) PDOS with LSDA+U (Anisimov *et al.*, 1996, 1997). A, tetrahedral coordinated Fe ions; B, octahedral Fe ions (B_1 corresponds to Fe^{3+} and B_2 to Fe^{2+} ions). The Fermi level (E_F) is set at 0.

The U value used by Anisimov *et al.* was 4.51 eV. The partial DOS for the two approximations, LSDA and LSDA+U, are compared in Figs. 4.35(a) and (b). In both cases, the O p-states are well separated in the region of -8 to -4 eV with respect to E_F. Thus, the results indicate the p-d hybridization between the O and Fe atoms is not strong. In the ↑-spin channel, the five d-bands of the Fe atom at the B site are occupied with the e_g states above the t_{2g} states. The d-states of the Fe at the A site form the conduction bands with the e_g state just above E_F. For the ↓-spin states, the role of the A and B sites is reversed. Furthermore, the t_{2g} states at the B site form the bottom of the conduction band. Since the calculations were intended to study charge ordering, the B sites were divided into B_1 and B_2 sites for Fe^{3+} and Fe^{2+}, respectively. At the B_1 site, $n^{B_1} = n_0 + \delta n$ and B_2 has $n^{B_2} = n_0 - \delta n$. The LSDA+U DOS shows that the t_{2g} states of the B_2 site in the ↓-spin channel experience a d–d interaction so that they split into bonding and antibonding states. All the other d-states, including states at the B_1 site, are unoccupied. Furthermore, the t_{2g} states of Fe at the B_1 site have energy below the antibonding states.

The LSDA predicts that Fe_3O_4 should be an HM with conducting states from the ↓-spin channel. On the other hand, in the LSDA+U calculations, the DOS shows that E_F touches the top of the valence band for both spin channels. The effect of U is to push up the ↑-spin e_g states at the B site (Fig. 4.35(a)) and lower the ↓-spin t_{2g} states at the B_2 site. The top edge of the t_{2g} states almost aligns with the e_g states.

Jeng *et al.* (2004) calculated the band structures of both crystal structures above and below T_V. For the low temperature case, they included U

Table 4.6. The U and J values used by Jeng *et al.* (2004).

Approximation	U (eV)	J (eV)
GGA	5.0	0.89
	4.5	0.5
	4.5	0.7
	4.5	1.1
LSDA	4.5	0.89
	5.0	0.89
	5.5	0.89

Table 4.7. The integrated charge in a sphere of radius 1.0 Å and the calculated M for the four Fe atoms.

Atom	Charge (e) Exp. (Wright *et al.*, 2001)	Theory	M (μ_B) Theory
Fe(B_1)	5.6	5.57	3.45
Fe(B_2)	5.4	5.41	3.90
Fe(B_3)	5.4	5.44	3.81
Fe(B_4)	5.6	5.58	3.39

in both LSDA and GGA exchange-correlation. The U and J values used by these authors are given in Table 4.6.

The magnitudes of U are comparable to that used by Anisimov *et al.* (1996). The results of these calculations are:

- With the refined low-temperature structure of Wright *et al.* (2001), the calculated results show not only the charge ordering consistent with the results of Wright *et al.* (2001) but also the orbital ordering.

- The integrated charge on each Fe atom is insensitive to the choice of U and J. The integrated charge in a sphere of radius 1.0 Å around each Fe atom is listed in Table 4.7. The calculated charges agree well with the measured values.

- The total energy in LSDA+U for the monoclinic (low temperature) structure is lower than the cubic structure by 0.35 eV/formula unit.

- Both GGA without U applied to the distorted lattice and LSDA+U applied to the undistorted cubic structure give metallic band structures in the minority-spin channel and a gap in the majority-spin states.

- The values of M listed in Table 4.7 are correlated to the charge states of the atoms. For Fe (B_1) and Fe (B_4) there is extra charge at the two sites causing a reduction of the net spin moment at these sites.

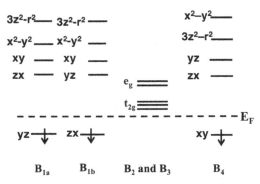

Fig. 4.36.   Schematic energy level diagram for the Fe ↓-spin d-orbitals under distortion at B sites in Fe_3O_4 (Jeng *et al.*, 2004).

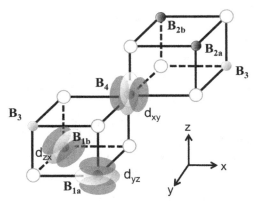

Fig. 4.37.   Orbital ordering in B-size sublattice. d_{yz} and d_{zx} orbitals are associated with the B_1 atoms, and d_{xy} orbitals are associated with the B_4 atoms (Jeng *et al.*, 2004).

- The charge ordering is explained by the lowering of one of the t_{2g} states of the Fe atoms at the B_1 and B_4 sites due to distortion associated with the Verwey transition. A schematic energy level diagram of the occupied and unoccupied states under the distortion is shown in Fig. 4.36. At the B_1 sites, the orbitals are d_{yz} and d_{zx}, respectively, while the one at the B_4 site is d_{xy}. They are now occupied and are depicted in Fig. 4.37. The shift of the orbitals is called orbital ordering and it is the driver of the charge ordering.

  The disagreement between the observed charge ordering (Wright *et al.*, 2001) and Anderson's pattern (Anderson, 1956) — minimal electrostatic repulsion — is explained. Anderson's pattern requires each tetrahedron with a corner sharing B type Fe to have two Fe^{2+} (B_1 and B_4) and two Fe^{3+} (B_2 and B_3) ions. This follows from Anderson's use of point charges. In reality, the charges are not point charges. Therefore, the screening effect is different.

- To obtain the charge-ordered state in agreement with the experimental results of Wright *et al.* (2001) does not require an intersite Coulomb interaction as introduced by Anisimov *et al.* (1996). It is also unnecessary to use different Us for Fe^{2+} and Fe^{3+} (Antonov *et al.*, 2001).

4.4. La_{1-x}(Sr, Ca, Ba)$_x$MnO$_3$

$La_{1-x}Sr_xMnO_3$ (LSMO) is considered as an alloy of $LaMnO_3$ with a La atom substituted by a Sr atom. $LaMnO_3$ is a perovskite-type oxide. It shows colossal MR and exhibits half-metallic properties for $x = 0.3$ (Park *et al.*, 1998b). At $x = 0.3$, it has been shown that its spin polarization is 95% at 4 K (Bowen *et al.*, 2003) and $\sim$90% at 100 K (Bowen *et al.*, 2005). Experimentally, as early as 1997, Wei *et al.* provided evidence that $La_{1-x}Ca_xMnO_3$ (LCMO) is an HM. $La_{2/3}Ba_{1/3}MnO_3$ (LBMO) shows giant negative MR (von Helmolt *et al.*, 1993). These alloys are potential materials for spintronic applications.

4.4.1. *Structure*

The crystal structure of perovskite is shown in Fig. 4.38. The Mn atom is surrounded by six O atoms forming an octahedron. This octahedron is the backbone of the FM oxides. LSMO exhibits a structural phase transition at $T_S = 280$ K with zero external magnetic field. T_S is a function of the external magnetic field and decreases to 220 K at magnetic field 70 kG. At low temperature, LSMO has the orthorhombic structure with space group Pnma. The high-temperature phase has the rhombohedral structure and belongs to the R3c space group. All these lattice parameters and angles depend on the value of x. For example, at $x = 0.17$, the lattice constants are $a = 5.547$ Å, $b = 7.790$ Å, and $c = 5.502$ Å. The structure consists of six formula units — six perovskite structures. Three of the octahedra stack up vertically and two of them align in the a-b plane. For the rhombohedral phase, the lattice constant is 5.475 Å and the angle α is 60.997°. Paiva-Santos *et al.* (2002) refined earlier crystal parameters of the $La_{0.65}Sr_{0.35}MnO_3$ rhombohedral structure determined by Alonso *et al.* (1997) using the Rietveld method (Rietveld, 1969). The method is now implemented in a software

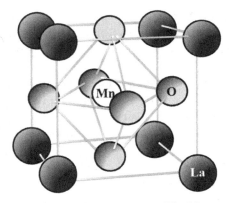

Fig. 4.38.   The unit cell of the perovskite structure. The Mn atoms (open circle at the center) occupy the A sites and the La atoms (black circles at corners of the cube) occupy the B sites. Oxygen atoms are shown as gray circles at face centers of the cube.

Table 4.8.   Refined inter-atomic distances between TM elements and an O atom, and the distance to the Mn atom in $La_{0.65}Sr_{0.35}MnO_3$.

TM element	Distance (Å)
La or Sr above and below the O atom	2.746(4)
La or Sr nearly in the same plane of the O atom	2.524(4)
Mn	1.954(4)

package (Young *et al.*, 1995). The lattice constant a is 5.5032 Å and c is 13.3675 Å. The value of a agrees reasonably well with the one obtained by Asamitsu *et al.* (1996). The interatomic distances of La or Sr to the O and Mn atoms are summarized in Table 4.8.

In fact, in $La_{1-x}Sr_xMnO_3$ the structural transition between the orthorhombic (O) and rhombohedral (R) phases depends on the temperature and Sr concentration x. The phase diagram expressed in terms of temperature and x is shown in Fig. 4.39, obtained from the neutron scattering experiments (Kawano *et al.*, 1996b). There are three sets of data points. The circles and squares mark the onsets of AFM and FM to paramagnetic phase transitions. The filled and open diamonds indicate the $R \rightarrow O$ transition temperature. The triangles indicate the $R \rightarrow O$ structural phase transitions for $x = 0.125$.

The neutron scattering results show also that there are two O phases, O$'$ and O*. The difference between these two phases was examined based on

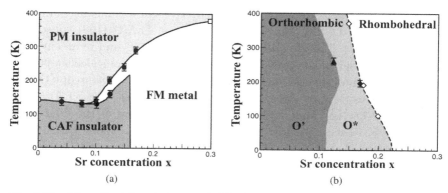

Fig. 4.39. Magnetic (a) and structural (b) phase diagrams of $La_{1-x}Sr_xMnO_3$ with $x \leq 0.17$. CAF is canted AFM and PM is paramagnetic. The circles and squares indicate the onset of FM and AFM components, respectively. The diamonds indicate the R $\rightarrow$ O transition temperature. The triangles denote the structural phase transitions for $x = 0.125$ (Kawano *et al.*, 1996b).

the powder pattern using Rietveld analysis. The O$'$ phase has a distorted MnO_6 octahedra due to the Jahn–Teller effect and its lattice constants satisfy $b/\sqrt{2} < c < a$. The O* phase exhibits the pseudocubic structure with $b/\sqrt{2} \sim c \sim a$ and without any evidence of the Jahn–Teller effect.

4.4.2. *Growth and characterization*

Depending on the properties to be probed, the growth involves bulk samples, polycrystalline forms, and thin films. Bulk $La_{1-x}Sr_xMnO_3$, $x = 0.35$ samples were grown by firing stoichiometric mixtures of Mn_2O_3, La_2O_3, and SrO_3 at $1300\,^\circ$C for 16 hours with intermediate grinding (Paiva-Santos *et al.*, 2002). X-ray diffraction was used to characterize the crystals and probe the possible refinement. Polycrystalline samples of LCMO were grown in two ways. One was to first grow powdered forms using similar mixtures as described above. The mixture was calcined three times in the air at $1050\,^\circ$C for 24 hours with intermittent grinding. There is a slight difference in detailed settings for the temperature and time (Asamitsu *et al.*, 1996; Kawano *et al.*, 1996a). The resulting powders were pressed into rods under a hydrostatic pressure of $1.4\,\text{ton/cm}^2$ then fired at $1100\,^\circ$C in air for 38 hours. Both groups used X-ray powder diffraction to characterize the samples. Another way to grow LCMO is to grow single crystals using the floating zone method. The initial samples were prepared in rod form and were loaded into a flow zone furnace. The feeding speed is 5–10 mm/hr. The single-crystal samples were then powdered. To form a tunneling junction, the thin films of LCMO were grown by PLD on $LaAlO_3(100)$ substrates (Wei *et al.*, 1997). The growth temperature was 700 °C and in an environment of 13.3 mbar of oxygen. The samples were then annealed at 900 °C in 1 atm oxygen for several hours. X-ray photoelectron spectroscopy (XPS) and STM were used to check the quality of the films. The XPS results showed very low contamination, indicating the surfaces of the films were clean. The STM was operated at 1 nA constant-current mode with Pt tip which was biased at about −2.0 eV. The images were taken at RT and 77 K in either high vacuum or ultra-pure He gas. They show atomically smooth "rice-paddy" type terraces with the step height of one unit cell.

4.4.3. *Physical properties*

Among the physical properties, half-metallicity is the prime interest with respect to the spintronic applications. In addition, magnetic and transport properties have attracted much attention. A recent review was given by Salamon and Jaime (2001).

4.4.3.1. *Half-metallicity*

Both LCMO and $La_{0.5}Ba_{0.5}MnO_3$ were predicted to be half-metallic by *ab initio* band structure calculations (Hamada *et al.*, 1995; Satpathy and Vukajlović, 1996). The first experimental evidence on LSMO was given by Okimoto *et al.* (1995) using optical conductivity measurements for $x = 0.175$, and by Hwang *et al.* (1996) measuring the MR for $x = 1/3$. Wei *et al.* (1997) carried out TJ measurements for LCMO with $x = 0.3$. The measured tunneling conductances for LCMO with $x = 1/3$ at $T = 77$ K (solid-dot curve) and 300 K (open circles) are shown in Fig. 4.40 for cases below and above T_C. The normalized conductances in Fig. 4.40 are proportional to the DOS. In practice, it normalizes the inherent dependence of the STM transmission probability on the voltage.

The important features are the two peaks at ±1.75 eV that appear in the low-temperature results. These two peaks agree with most of the DOS derived from the spin-polarized band structures of LSMO. The DOS results exhibit the half-metallic properties for LSMO (Singh and Pickett, 1998; Ma *et al.*, 2006). The −1.75 eV peak corresponds to the peak in the occupied states of the majority-spin channel and the +1.75 eV peak corresponds to the peak in the unoccupied minority-spin states above E_F. The difference is the strength of the exchange splitting. The half-metallic feature is explicitly

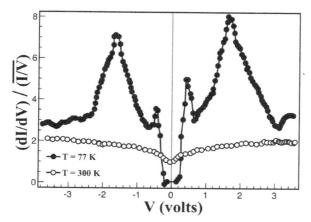

Fig. 4.40. Normalized conductance as a function of bias voltage of LCMO with $x = 1/3$, at $T = 77$ K (solid-dot curve) and $T = 300$ K (open circles) (Wei *et al.*, 1997).

shown by the presence of the spin splitting peaks in the low-temperature results. At 300 K, there are no such peaks consistent with the paramagnetic properties.

4.4.3.2. *Magnetic and transport properties*

T_C is the primary interest regarding the magnetic properties of the oxide alloys involving La. The magnetization and hysteresis loops will also be discussed. As for the transport properties, we shall discuss primarily the various MRs.

Magnetic properties T_C in these oxide alloys varies with respect to the composition x. Determination of T_C as a function of x for LSMO was carried out by Jonker and Santen (1950). Table 4.9 lists T_C for several of these oxides for a few x compositions. Shown in Fig. 4.41 are the magnetization $M(T)$ as a function of temperature and hysteresis loop of LCMO at $x = 0.3$ (Park *et al.*, 1998b). The sample was epitaxially grown in film form on a $SrTiO_3(001)$ substrate by PLD. $M(T)$ was measured by SQUID and T_C is extrapolated to be 360 K. The hysteresis loop was determined by MOKE. It shows that the sample has a single domain as indicated by nearly 100% remanent M and very low coercive field H_{co}. For bulk samples, H_{co} is approximately 30–50 Oe and M is saturated at 5 kG (McCormack *et al.*, 1994).

Table 4.9.   Summary of T_C for the alloying oxides.

Alloy	x	T_C (K)	Reference
LSMO	0.17	264	Asamitsu *et al.* (1996)
	0.175	283	Okimoto *et al.* (1995)
	0.3	360	Park *et al.* (1998b)
	1/3	370	Bowen *et al.* (2003)
	0.3–0.5	~380	Asamitsu *et al.* (1996)
LCMO (film)	0.3	260	Wei *et al.* (1997)
LaBa (LB) MO	0.33	343	von Helmolt *et al.* (1993)

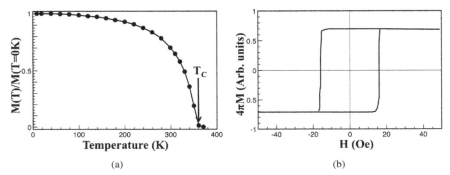

(a)   (b)

Fig. 4.41.   (a) $M(T)/M(T = 0 \text{ K})$ as a function of temperature and (b) a hysteresis loop at $T = 300$ K for LCMO with $x = 0.3$ (Park *et al.*, 1998b).

Transport properties   Experiments providing evidence for LBMO with $x = 1/3$ exhibiting giant negative MR were carried out by von Helmolt *et al.* (1993). The samples were films in thickness of 150 ± 10 nm grown by PLD on (100)- and (110)-oriented $SrTiO_3$ (STO) substrates. The unit cell is rhombohedral. MR is defined as MR $= [R(0) - R(50\text{kG})]/R(0)$. At RT, MR is more than 60% larger than that of Cu/Co multilayer structures. In metallic multilayers, MR is attributed to the spin-dependent scattering at the interface between the magnetic and nonmagnetic layers. The MR in LBMO is independent of the relative orientation of H, current I, and axis of the samples. The high MR and formation of partially localized states at the magnetic impurities were the basic reasons for von Helmolt *et al.* to suggest that the carriers contributing to the transport are magnetic polarons.

TMR and inverse TMR measurements were carried out by Lu *et al.* (1996) and De Teresa *et al.* (1999), respectively. The TMR ratio is $\Delta R/R = (R_{\uparrow\downarrow} - R_{\uparrow\uparrow})/R_{\uparrow\downarrow}$, where $R_{\uparrow\downarrow}$ and $R_{\uparrow\uparrow}$ are resistances for the two sides of the junction in antiparallel and parallel configurations. The trilayer sample of

LSMO/STO/LSMO was grown by PLD. The two LSMO layers are approximately 500 Å thick and the barrier is 30–60 Å. The MR value of 83% was obtained. From the result, the value of P deduced from the Jullière formula was 54%, which is in good agreement with the value of 53% predicted by Singh and Pickett (1998) for $La_{2/3}Ba_{1/3}MnO_3$. The inverse TMR measurement was carried out with trilayers of LSMO ($x = 0.3$)/STO/Co. The thicknesses of the three layers were 35/2.5/30 nm, also grown by PLD. In this layer configuration, the LSMO is used as a spin analyzer for the polarization in Co. Experiments were carried out at $T = 5$ K and –50% inverse TMR was measured. The "inverse" is defined as the negative TMR, having small resistance in the antiparallel configuration. The carriers originate from the negatively polarized electrons in the Co layer. It is well known that the DOS of Co at E_F for the majority-spin channel is smaller than the corresponding DOS of the minority-spin states. In Fig. 4.42, the data for the MR exhibit the inverse TMR. A qualitative explanation of the MR results is given in Fig. 4.43. The DOS in the minority-spin channel of Co at E_F is larger than that of LSMO. At bias voltage $V = 0$V (Fig. 4.43(a)), the current flows from the left (majority) to the right (minority). From the TMR ratio, $R_{\uparrow\uparrow}$ is larger. When V is in the range of 0.7 and -2 V (Fig. 4.43(b)), E_F in the Co region is lowered and the current flows from the left to the right because the relative feature of the DOS for LSMO and Co is not changed. The current reverses its direction when V is greater

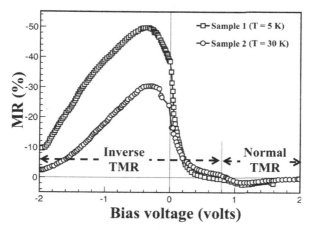

Fig. 4.42. MR as a function of bias voltage of two LSMO/STO/Co samples at $T = 5$ K and 30 K (De Teresa *et al.*, 1999).

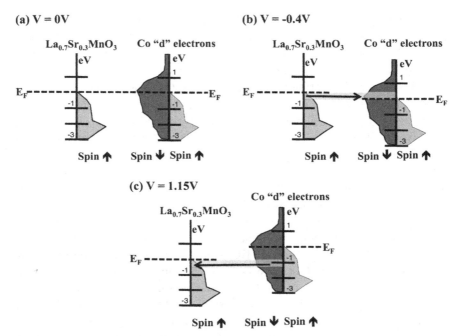

Fig. 4.43. Schematic diagram of the tunneling processes corresponding to MR in Fig. 4.42. The vertical axis in each panel is the energy, the horizontal axis is the DOS, and E_F is indicated by dashed lines. The thick arrows indicate the direction of current flows (De Teresa *et al.*, 1999).

than $1.15\,$V because $R_{\uparrow\downarrow}$ is larger because the DOS of the Co majority-spin states is now larger (Fig. 4.43(b)).

More recently, the spin-polarized TMR in LSMO (350 Å)/STO (001) (28 Å)/LSMO (100 Å) with $x = 0.3$ was determined by Bowen *et al.* (2003). These trilayers differ from the earlier samples in three ways:

- Higher-quality interfaces of LSMO/STO.
- Smaller sizes of junctions.
- Pinning the spin of the top layer of LSMO. The pinning was accomplished as follows: the samples were first inserted into a radio frequency sputtering system and were covered by a 150 Å Co layer on top. Then, they were etched by an oxygen-rich plasma to form a CoO layer of 25 Å in thickness. Finally, the samples were capped by a 150 Å Au layer. The TMR is 1800% at $T = 4.2\,$K under a bias voltage of $1\,$mV. The deduced P is 95%. The TMR vanishes at $T = 280$ K. The improved TMR was attributed to the sample preparation.

One-electron properties In this section, we discuss one-electron theory, in particular band structures. Effects of strains, the Jahn–Teller distortion, and rotations of spin moments on some Mn atoms are the primary issues. Most of the calculations were done using LSDA with the LAPW method as used by Singh and Pickett (1998). The FLEUR code,[4] which has the spin polarization and other modern features, is a popular implementation, as used by Ma *et al.* (2006).

Singh and Pickett have studied effects of the rotation of spin moments and the Jahn–Teller distortion in $Mn-O_6$ octahedra. Similarly, Ma *et al.* (2006) also used a 20-atom model with Pnma symmetry and a 5-atom perovskite unit cell. The model is shown in Fig. 4.44. The lattice constants and MT radii used by Ma *et al.* are summarized in Table 4.10.

The conclusions of their study are summarized as follows:

- Mn(1) has 3.11 μ_B moment and Mn(2) has 3.26 μ_B.
- Using GGA exchange-correlation, the total magnetic moment/unit cell is 10.89 μ_B. With GGA+U, the moment is 11.0 μ_B. The half-metallicity of LSMO with $x = 1/3$ is therefore sensitive to the choice of U value used in the calculations.
- The effects on the VBM (valence band maximum) and CBM (conduction band minimum) due to the in-plane compression were examined with and without U. For $U = 0$, the VBM is higher and the CBM is

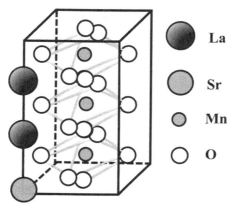

Fig. 4.44. A 20-atom model of LSMO (Ma *et al.*, 2006). Sr is shown as a large gray circle, filled black circles are La, small gray circles are Mn, and open circles are O.

---

[4]http://www.flapw.de

Table 4.10. Summary of lattice constants and MT radii of LSMO with $x = 1/3$ used in LAPW calculations (Ma *et al.*, 2006).

Lattice constant (Å)		MT radius (Å)				U (eV)	J (eV)
a	c	La	Sr	Mn	O		
3.90	$3a$	1.32	1.32	1.06	0.794	2 and 3	0.7

lower with respect to E_F by a few tenths of an eV. With $U = 2.0\,\text{eV}$, the VBM is essentially unchanged while the CBM shifts down by about the same amount as for $U = 0$. Consequently, the effects of the strain are not drastic.

- The occupied states near E_F are the e_g states. The in-plane strain lifts the degeneracy of $x^2 - y^2$ and $3z^2 - r^2$ states.
- The origin of the magnetism was attributed to the mixed valence of the Mn ions, Mn^{3+} and Mn^{4+}, and an electron hopping between them.

In treating the LSMO alloy within the DFT calculations, Singh and Pickett (1998) used virtual crystal approximation for $La_{2/3}Ba_{1/3}MnO_3$. The size of the unit cell is that of LCMO with $x = 1/3$ exhibiting colossal MR (CMR). Relaxations were performed in FM and AFM configurations. Their results are summarized as follows:

- The spin moment on the Mn atom is $3.40\,\mu_B$.
- The DOS does not show any half-metallic property.
- The FM phase has a lower energy than the AFM phase by $0.014\,\text{eV/Mn}$.
- The rotations of the octahedra affect the e_g states which form the gap near E_F.
- The lattice distortions in FM alloys do not couple to electrons as strong as in the spin-disordered paramagnetic state.

4.5. Magnetic Interactions in the Oxides

Understanding the physical origins of the magnetic properties in the oxides remains a challenging task. We shall first consider the basic magnetic interactions discussed in the literature. Then, we discuss some recent results. The key interactions are:

(1) Superexchange interaction.
(2) Double exchange interaction.

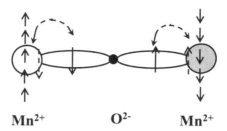

Mn^{2+} O^{2-} Mn^{2+}

Fig. 4.45. Superexchange interaction. The dashed arrows indicate that the electrons can hop.

4.5.1. *Superexchange*

4.5.1.1. *Model approach*

First proposed by Kramers (1934), it was Anderson (1950) who simplified Kramers' general formalism and provided a simple picture. The effective interaction based on Kramers' formulation between two Mn atoms with a common neighboring O atom is called superexchange. The essential idea of the interaction is shown in Fig. 4.45. The O^{2-} ion is located between two Mn^{2+} ions. There are no free carriers in Anderson's view of superexchange.

The O^{2-} ion has a filled p-shell so that it has a net charge of $-2e$. The two electrons occupy one of the p-orbitals with spins pointing in opposite directions. The Mn^{2+} ions have five d-electrons and align their local spins according to Hund's first rule. O p-electrons can hop to the neighboring unoccupied Mn d-states while conserving spin. This delocalization reduces the energy of the O^{2-} ion. Because the spins of p-electrons are oppositely oriented, this hopping can only occur if the neighboring d electrons are antiferromagnetically aligned. The effective interaction was reformulated in terms of spin operators and the orthogonality of the wave functions.

In reality, the wave functions at the neighboring atoms are not orthogonal. This led Yamashita and Kondo (1958) to reconsider the superexchange interaction. These authors pointed out that if the ionic model shown in Fig. 4.45 is considered, then it is necessary to expand the total energy to 4th order in the overlap integral S in order to have the total energy dependent on the spin arrangements. They argued that the ionic states should include some mixing of electronic configurations for excited states and investigated the possible contributions from 4 different configurations:

- The Slater model (Slater, 1953). This model consists of $[Mn^{2+}; O^{2-} (2p)^6]$(the ionic model) $+ [Mn^{2+}; O^{2-} (2p\uparrow)^2 (3s\uparrow)(2p\downarrow)^3] + [Mn^{2+}; O^{2-} (2p\uparrow)^3 (3s\downarrow) (2p\downarrow)^2]$.

- The Kramer–Anderson model (Anderson, 1950). It has $[Mn^{2+}; O^{2-}] + [Mn^+; O^-]$.
- The Goodenough model (Goodenough, 1955). Two configurations are $[Mn^{2+}; O^{2-}] + [Mn^+—O—Mn^+]$.
- The band model (Yamashita and Kondo, 1958). The model also has two configurations, $[Mn^{2+}; O^{2-}] + [Mn^{3+}—O^{2-}—Mn^+]$.

The Kramer–Anderson model can have both FM and AFM states. All the other models give AFM states due to symmetry or the Pauli principle. Furthermore, both the Goodenough model and band model indicate that the Mn ions can have mixed valence. The important point is that it is the overlap matrix elements which contribute to superexchange.

4.5.1.2. *First-principles approach*

Oguchi *et al.* (1984) approached superexchange in TM oxides and sulfides such as MnO, NiO, and MnS, from the band structure point of view. They discuss the "itinerant versus localized picture of superexchange." The use of the word "itinerant" conflicts with Anderson's localized picture of superexchange. In electronic band structures, localized electrons have flat bands. The authors pointed out that the superexchange is associated with second neighbor AFM interactions characterized by the exchange constant J_2 originating from the e_g-p-e_g coupling. It is known that the e_g states in Mn form narrow bands. In this sense, it is consistent with the localized picture of Anderson. The approach has no empirical parameters and the interaction strength for a crystal is inversely proportional to the exchange splitting instead of the U term in the model approach. Qualitatively, this explains the increasing trend of the Néel temperatures from MnO to NiO.

4.5.2. *Double exchange*

Zener (1951) proposed a mechanism to explain the FM properties of oxides, now known as the double exchange. The double exchange interaction is like that of superexchange in that it causes alignment of neighboring magnetic ions to minimize the kinetic energy of hopping electrons. However, in double exchange the alignment is FM rather than AFM as in superexchange. It typically occurs in systems with mixed valence magnetic ions, such as Fe_3O_4. This is because such systems can have neighboring magnetic ions of the same species but different valences: for example, the Fe^{2+} and Fe^{3+} ions at the octahedral B sites in Fe_3O_4. In such a case, hopping of the unpaired d-electron of the Fe^{2+} ($3d^6$) to the Fe^{3+} ($3d^5$) can occur only if the remaining d-electrons are aligned ferromagnetically. Thus kinetic energy is reduced by this configuration. The interaction can be understood in terms of two degenerate configurations, Fe^{2+}—O^{2-}—Fe^{3+} and Fe^{3+}—O^{2-}—Fe^{2+}. They are connected by double exchange matrix elements characterizing an electron hopping from Fe^{2+} to O^{2-} and from O^{2-} to Fe^{3+}. It is important to realize that the transfer matrix element in this case is nonzero only when the local spins at Fe sites are aligned according to Hund's first rule. There is a resonance of local spins in neighboring Fe atoms if the spins are parallel. The splitting due to double exchange matrix elements is of the order of $k_B T$, where k_B is the Boltzmann constant.

Anderson and Hasegawa (1955) revisited Zener's theory (Zener, 1951) using a model for treating *classically* the localized spins at the Mn ions and *quantum mechanically* the spin of a mobile electron characterized by the double exchange matrix elements. A simple picture is shown in Fig. 4.46. The Mn ions have their d-electron spins aligned according to Hund's first rule. The spin moments for the left and right Mn ions are S_1 and S_2, respectively. The right Mn ion has a mobile electron with spin moment s. Three cases listed in Table 4.11 are considered. The important quantities are the intra-atomic exchange integral J between mobile electron and local d-orbitals, and the magnitude of the transfer matrix element b characterizing the effective coupling between Case 1 and Case 2. The authors showed that by coupling states from two Mn ions with b, the splitting of the Zener levels is proportional to $\cos(\frac{\theta}{2})$, where θ is the angle between classical spins S_1 and S_2 at the two Mn ion sites and the effective transfer integral is

$$t_{\text{eff}} = b \cos\left(\frac{\theta}{2}\right) = b\frac{(S_0 + \frac{1}{2})}{2s + 1}, \tag{4.18}$$

where S_0 is the sum of the spin moments of the two magnetic ions, and s is the spin of the mobile electron. With $b > J$, they showed that the FM

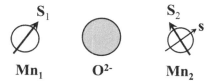

Fig. 4.46. Double exchange interaction. S_1 and S_2 are the spin moments of the Mn ions contributed by the localized d-electrons. s is the spin of the itinerant electron.

Table 4.11.   Three cases considered in double exchange. d_1 and d_2 are the localized d-states contributing to $\mathbf{S}_1$ and $\mathbf{S}_2$. d is the state of the mobile electron and p^2 and p are the doubly and singly occupied states of the O atom.

Case	Mn$_1$	O	Mn$_2$
1	d_1	p^2	d_2, d
2	d_1, d	p^2	d_2
3	d_1, d	p	d_2, d

phase can occur through the mobile electron. It is possible to conclude that double exchange favoring ferromagnetism should involve:

- Localized spin at each TM ion site.
- Hund's first rule coupling the localized spin and spin of the mobile electron.

Consequently, the motion of the mobile electron between the two sites is correlated. Most of the investigations of the magnetic properties of the metal oxides have been based on either superexchange or double exchange or both, depending on the particular oxide of interest.

4.5.3.   Magnetism in CrO_2 and Fe_3O_4

There has been much work investigating the magnetic interactions in CrO_2 and Fe_3O_4. Because much has been learned about these two important cases, we consider each in turn below.

4.5.3.1. CrO_2

For CrO_2, there is general consensus that it is the double exchange mechanism that is responsible for its magnetic properties. Schlottmann (2003) combined single-particle results with collective excitations to show that double exchange in CrO_2 depends critically on the distortion of the octahedron surrounding the Cr atom and therefore differs from the double exchange mechanism in the manganites.

The physical picture is as follows: the Cr ion in the octahedron formed by the O atoms has valence 4+. Only two d-electrons remain in valence. Due to the cubic environment, the five d-states split into t_{2g} (triply degenerate) and e_g (doubly degenerate) states, with the t_{2g} states at lower energy. Therefore, the t_{2g} states are partially occupied and e_g states are unfilled.

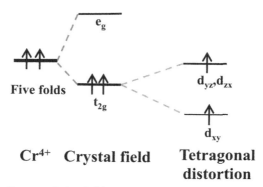

Cr^{4+} Crystal field   Tetragonal distortion

Fig. 4.47.   Level diagram of the Cr^{4+} ion under cubic crystal field and tetragonal distortion. One of the two electrons occupying the d$_{xy}$ state is localized while the other electron sharing $\frac{1}{\sqrt{2}}$ (d$_{yz}$ ± d$_{zx}$) states is itinerant.

The t$_{2g}$ states are composed of d$_{xy}$, d$_{yz}$, and d$_{zx}$ orbitals. The octahedron undergoes a Jahn–Teller distortion. This tetrahedral distortion lifts the degeneracy of the t$_{2g}$ states with the d$_{xy}$ state having the lowest energy. The other two orbitals form $\frac{1}{\sqrt{2}}$ (d$_{yz}$ + d$_{zx}$) and $\frac{1}{\sqrt{2}}$ (d$_{yz}$ − d$_{zx}$) states. One of the two electrons occupies the d$_{xy}$ state and the other electron has a 50% probability of occupying each of the combined orbitals. The level scheme is shown in Fig. 4.47.

To simplify the model for analysis, the following approximations are made:

- Direct hopping between the Cr atoms rather than mediation by the O p-states.
- The localized state at each site does not participate in hopping.
- Hopping between the same orbital angular momentum states.

Letting the hopping integral between the two sites, 1 and 2, be t, the hopping Hamiltonian is

$$H_t = -t \sum_\sigma \left(c^\dagger_{1\xi\sigma} c_{2\xi\sigma} + c^\dagger_{2\xi\sigma} c_{1\xi\sigma} + c^\dagger_{1\zeta\sigma} c_{2\zeta\sigma} + c^\dagger_{2\zeta\sigma} c_{1\zeta\sigma} \right), \qquad (4.19)$$

where ζ and ξ label orbital angular momentum states and σ labels the spin.

The other term required is the on-site Coulomb energy U to model the full Hamiltonian. To apply the model Hamiltonian to CrO$_2$, we note that there are two Cr ion sites, with two electrons each. One of them occupies the d$_{xy}$ localized state and the other is itinerant. From Hund's first rule at each site, the orbital wave function is odd under permutation of the two electrons and the spin state is triplet. Combining the two sites, the resultant spin momenta, $S = 2, 1, 0$, can be constructed and serve as the states acted on by H_t. The $S = 2$ case consists of three states having an orbital triplet and one state having an orbital singlet. The triplet is even under permutation of two mobile electrons while the singlet is odd.

$$\psi_{even}^{\zeta\zeta}(S = 2) = c_{1\zeta\uparrow}^{\dagger}c_{2\zeta\uparrow}^{\dagger}|1\uparrow 2\uparrow\rangle$$

$$\psi_{even}^{\zeta\xi}(S = 2) = \frac{1}{\sqrt{2}}(c_{1\zeta\uparrow}^{\dagger}c_{2\xi\uparrow}^{\dagger} + c_{1\xi\uparrow}^{\dagger}c_{2\zeta\uparrow}^{\dagger})|1\uparrow 2\uparrow\rangle \tag{4.20}$$

$$\psi_{even}^{\xi\xi}(S = 2) = c_{1\xi\uparrow}^{\dagger}c_{2\xi\uparrow}^{\dagger}|1\uparrow 2\uparrow\rangle$$

$$\psi_{odd}^{\zeta\xi}(S = 2) = \frac{1}{\sqrt{2}}(c_{1\zeta\uparrow}^{\dagger}c_{2\xi\uparrow}^{\dagger} - c_{1\xi\uparrow}^{\dagger}c_{2\zeta\uparrow}^{\dagger})|1\uparrow 2\uparrow\rangle, \tag{4.21}$$

where $c_{1\xi\uparrow}^{\dagger}$ is the creation operator of an electron at site 1 occupying state ξ — one of the linear combinations of d_{yz} and d_{zx} with $\uparrow$ spin. The even states are unchanged under H_t. The energy of the odd state is lowered by $4t^2/U$. This is the largest shift among all values of S. Therefore, FM ordering is favored, consistent with the double exchange mechanism.

4.5.3.2. Fe_3O_4

The Fe ions in Fe_3O_4 have two different valences, $2+$ at the B_2 and B_3 sites and $3+$ at the B_1 and B_4 sites. In addition, the Verwey transition complicates investigations of the origin of the ferromagnetism in this compound. At present, the origins of the ferromagnetism in Fe_3O_4 and the La oxide alloys are still somewhat controversial. In this section, we discuss a treatment considering the double exchange mechanism but taking the AFM ordering between A and B sites into account. The approach differs from that of the preceding section, considering instead the free energy and minimizing it with respect to the magnetization at the B site (Loos and Novák, 2002).

The double exchange mechanism proposed by Loos and Novák (2002) was based on the quantum mechanical formulation of Kubo and Ohata (1972). Its original application was to La oxide alloys. Grave $et~al.$ (1993) applied a modified formulation (Kubo and Ohata, 1972) to Fe_3O_4. In the original formulation for La oxides alloys, the picture is that the Mn^{3+} has three electrons occupying the t_{2g} manifold and one electron in the higher energy e_g states. The e_g states on the Mn^{4+} are unoccupied. The electron in the e_g states of Mn^{3+} is considered as itinerant. It can hop from Mn^{3+} to

Mn^{4+} and vice versa. The assumptions are that the intra-atomic exchange integral J is larger than the hopping integral t_{ij}, $J \gg |t|$, the charge effect due to alloying is averaged out, and the alloys can be treated within the virtual crystal approximation. The Hamiltonian characterizing the double exchange is

$$H = -J \sum_{i,s,s'} \left(S_i \sigma_{ss'} c_{is}^\dagger c_{is'} \right) + \sum_{i,j,s,s'} \left(t_{ij} c_{is}^\dagger c_{js'} \right), \qquad (4.22)$$

where i and j are site indices, S_t is the total spin of the localized state at site i, $\sigma_{ss'}$ is the Pauli matrix, and the c's are the creation and annihilation operators for the itinerant electron. Recall that in the spinel structure of Fe_3O_4 shown in Fig. 4.16, there are tetrahedral (A) sites and octahedral (B) siets. The B sites contribute to the local spins and itinerant electrons (Loos and Novák, 2002). The mean-field approach was adopted to find an effective field acting on an atom at a B site due to its neighbors.

Above the Verwey temperature, there are N_B sites with each B site having a local spin moment S of $(5/2)\hbar$ and $N_B/2$ itinerant electrons. Let S' be the effective local spin moment resulting from the spin moment of the itinerant electron coupled to the local spin through Hund's first rule. Grave *et al.* (1993) obtained a value of 2.9 for Fe_3O_4 through their hyperfine measurements based on the double exchange formulation of Kubo and Ohata (1972). This differs quite significantly from the ideal estimation of 2.0. Loos and Novák (2002) addressed this discrepancy by including superexchange between site A and site B. This is done by using effective fields which are proportional to the z-components of the equilibrium magnetizations, $m_{A,\text{eq}} = \langle S_{A,z} \rangle_{\text{eq}}$ and $m_{B,\text{eq}} = \langle S_{B,z} \rangle_{\text{eq}}$, acting respectively on site B and site A ions. m_A, therefore, can be expressed as

$$m_A = SB_S(\lambda_{AB} m_{B,\text{eq}} + \lambda_{AA} m_A), \qquad (4.23)$$

where B_S is the Brillouin function. The λ's characterize the superexchange interactions, with

$$\lambda_{ij} = \beta_B S J_{ij} z_j(i), \qquad (4.24)$$

where J_{ij} is the strength of the superexchange interaction between sites i and j, $z_j(i)$ is the coordination number of the i-th atom with j nearest neighbors, and β_B is $\frac{1}{k_B T}$. Let λ, the mean field at T, be $\beta S g \mu_B H_{\text{eff}}^z$ acting at a B site; the magnetization at the same site is

$$m_B(\lambda) = \langle S_B^z(\lambda) \rangle = (1 - x)SB_S(\lambda) + xS'B_{S'}(S'\lambda/S), \qquad (4.25)$$

where g is the Landé g-factor and x is the fraction of itinerant electrons relative to the total at site B. S is the total local spin and S' is the resultant spin at site B including the local spin S and the spin of the itinerant electron coupled by Hund's rule (Loos and Novák, 2002). The effective Hamiltonian due to the superexchange and the free energy per B site f_B can then be given:

$$\langle H_{\mathrm{SE}} \rangle = -\frac{1}{\beta_B S} \left(m_B \lambda_{BA} m_{A,eq} + \lambda_{BB} \frac{m_B^2}{2} \right) \qquad (4.26)$$

$$f_B = x\mu + \Omega + \langle H_{\mathrm{SE}} \rangle - TS^{(S)}, \qquad (4.27)$$

where μ is the chemical potential for the itinerant electron. It is expressed in terms of the DOS, $g(\varepsilon)$, of the free carriers,

$$\mu = \int d\varepsilon \frac{g(\varepsilon)}{e^{\beta_B(\varepsilon-\mu)} + 1}, \qquad (4.28)$$

grand canonical potential

$$\Omega = \int d\varepsilon g(\varepsilon) \ln(e^{\beta_B(\varepsilon-\mu)} + 1), \qquad (4.29)$$

and entropy,

$$S^{(S)} = k_B \left\{ \left[(1-x) \ln \left(\sum_{|m| \leq \frac{5}{2}} e^{\lambda m/S} \right) \right. \right.$$

$$\left. \left. + x \ln \left(\sum_{|m| \leq 2} e^{\lambda m/S} \right) \right] - \frac{\lambda m_B}{S} \right\}. \qquad (4.30)$$

m_B can then be determined by minimizing f_B with respect to m_B. The theory has some shortcomings. If S' is set equal to 2, then T_C is 1590 K, almost a factor of 2 larger than experiment. Loos and Novák (2002) attributed this shortcoming to the mean field theory.

Chapter 5

Half-metals with Simple Structures

5.1. Introduction

Both Heusler alloys and TM oxides are appealing for spintronic applications. However, due to Coulomb correlation at the TM sites, defects, phase transitions (Borca *et al.*, 2001), and surface and interface effects in thin-film samples, neither Heusler alloys nor TM oxides have been shown experimentally to exhibit half-metallic properties at RT. Thus at the beginning of this century, researchers began to search for HMs with simpler structures, such as the ZB structure. Exciting results were obtained by Akinaga *et al.* (2000a). They first predicted CrAs in the ZB structure to be an HM by first-principles calculations then grew it in thin-film form. In this new HM, the unit cell consists of just two atoms, as in GaAs. It is especially appealing in light of the following:

- The simple structure can eliminate the disorder problem.
- It can be integrated readily into well-developed semiconductor technologies because its structure conforms to semiconductors in common use today.
- If single crystals or thin films can be grown, transport properties can be improved; this is in contrast to Heusler alloys which are more susceptible to defects.

As for most of the Heusler alloys, there is as yet no experimental evidence of half-metallicity for ZB compounds such as CrAs. The results of Akinaga *et al.* (2000a) have nevertheless stimulated much research to design new HMs with simple structures. In this chapter, we shall first discuss the efforts undertaken by various researchers to grow (Zhao *et al.*, 2001; Etgens *et al.*, 2004) and predict (Continenza *et al.*, 2001; Galanakis, 2002a; Xu *et al.*, 2002; Zhao *et al.*, 2002; Galanakis and Mavropoulos, 2003; Liu, 2003; Pask *et al.*, 2003; Xie *et al.*, 2003; Şaşıoğlu *et al.*, 2005a) new HMs in the ZB structure. Since the growth of these compounds is the major hurdle, due to the fact that the ZB structure is not the ground state structure (e.g., many have the hexagonal NiAs structure as the ground state, in which they are not half-metallic), we give the details of the growth techniques adopted by various researchers to provide a foundation for future developments. The interactions giving rise to the half-metallicity in these compounds will be discussed. We shall then describe the efforts to grow (Mizuguchi *et al.*, 2002; Akinaga and Mizuguchi, 2004) and design quantum structures, including superlattices (Fong *et al.*, 2004; Fong and Qian, 2004), quantum dots (Qian *et al.*, 2004a; Shirai, 2004), and digital FM heterostructures (DFH) (Sanvito and Hill, 2001; Qian *et al.*, 2006a; Wu *et al.*, 2007; Zhu *et al.*, 2008), and discuss their physical properties. Half-metallic Si-based DFHs in particular will be elaborated, due to the wide availability of mature Si-related technologies. Finally, we shall discuss attempts to design even simpler structures. In this respect, carbon nanowires doped with TM elements are a promising direction (Dag *et al.*, 2005; Durgun *et al.*, 2006).

5.2. Half-metals with Zincblende Structure

Two TM pnictides with ZB structure have been grown in thin-film form. All other TM pnictides, a carbide, and chalcogenides have been predicted by calculations.

5.2.1. *Experiment*

We shall focus on the experimental efforts on the growth, characterization, and magnetic properties of HMs with ZB structure.

5.2.1.1. *Growth*

There are two forms for the growth: thin film and multilayers.

Thin films of CrAs   In 2000, Akinaga *et al.* (2000a) used MBE to grow CrAs on a GaAs(100) substrate. A molybdenum holder with indium solder was used to hold the substrate. A 20 nm GaAs buffer layer was first grown after the chamber was degassed. The CrAs was then grown on top at 0.02 nm/s. The temperature of the growth was 200 °C. Finally, a 4.0 nm layer of gold was used to cap the CrAs for preventing oxidations. The resulting thickness of the CrAs layer with the ZB structure is 2 nm.

The results reported by Akinaga *et al.* were not without controversy. Etgens *et al.* (2004) investigated the structure of CrAs epilayers grown on

GaAs(001) and also used the MBE method to grow their samples. A 100 nm undoped GaAs buffer layer was first grown on GaAs(001) substrate. The surface was annealed under As to improve the flatness of the surface. CrAs was then grown at 200 °C with a rate of 8 Å/min under As-rich conditions. At 295 K, CrAs has the orthorhombic MnP structure with lattice constants, $a = 5.637$ Å, $b = 3.445$ Å, and $c = 6.197$ Å. Above 1100 K, it is grown in the hexagonal NiAs structure.

Thin films of CrSb Zhao *et al.* (2001) reported the growth of CrSb on GaSb. They also used the MBE method to grow CrSb films on a GaAs(001) substrate with three different buffer layers, GaAs, AlGaSb, and GaSb, respectively. The layer configurations are shown in Fig. 5.1. The solid sources of Ga, As, Sb, Al and Cr were provided. Before depositing the buffer layer, the oxide layer on top of the substrate was cleaned at 580 °C. The buffer layers have different thicknesses as shown in Fig. 5.1. For example, case A has 500 nm thick GaAs and was grown at 560 °C. During the growth on this buffer layer, the As and Ga fluxes were terminated simultaneously at the end of the growth. The Sb fluxes were immediately released to prevent the sublimation of group-V elements from the surface. At this time, the temperature of the substrate was lowered to grow CrSb. When the substrate temperature reached 400 °C, the flux of Sb atoms was stopped to avoid the build-up of Sb on the surface. As the temperature of the substrate reached 250 °C, both Cr and Sb fluxes were then turned on. Typically, the intensity of the Cr flux was 2×10^{13} cm^{-2} s^{-1}. The beam equivalent pressure ratio of Sb/Cr was maintained at about 10. By monitoring RHEED patterns that show ZB features in the sample, these authors

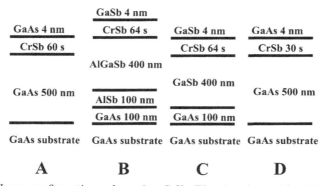

Fig. 5.1. Layer configurations of growing CrSb. The time (seconds) of deposition for the GaSb layers is also shown (Zhao *et al.*, 2001).

Table 5.1.   Summary of information for growing CrSb with the ZB structure on three different buffer layers.

Sample	Buffer layer	Growth time (s)	Duration of ZB RHEED pattern (s)	Substrate temp. (°C)	Cr flux $(10^{13}\,\mathrm{cm}^{-2}\,\mathrm{s}^{-1})$
A	GaAs	60	40	250	2
B	AlGaSb	64	60	250	3
C	GaSb	64	60	250	2

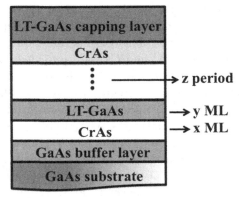

Fig. 5.2.   Schematic diagram showing the layer configurations, x, y, and z of CrAs/GaAs multilayers. LT means low temperature (220 °C) with respect to the buffer layer temperature (580 °C) (Mizuguchi *et al.*, 2002).

specified the thickness in terms of depositing time. Typically, the growth time was 60 s. The results are summarized in Table 5.1.

Multilayer growth   Recognizing that the ZB structure is not the ground state structure for CrAs and CrSb, the growth of multilayer films has been pursued by Mizuguchi *et al.* (2002). They used an MBE system to grow a CrAs/GaAs multilayer with period z on semi-insulating GaAs(001) substrates. A sample of the multilayer with a period z consisting of x layers of CrAs and y layers of GaAs is shown in Fig. 5.2.

We now discuss the growth of the multilayers. In the UHV chamber, thermal cleaning was carried out by annealing the substrate at 600 °C for 10 minutes to remove the surface oxidation layer on the GaAs substrate. A GaAs buffer layer of 20 nm thickness was grown at 580 °C. Then CrAs and GaAs layers were grown alternately. They were controlled by opening shutters of Knudsen cells for each pair of elements simultaneously at 220 °C.

The vapor pressure ratio of As/Cr was set at 100 to 1000. The period of the growth z was set at 10 or 100. The thickness of the last GaAs layers was set at 5 nm to prevent the oxidation of the multilayer structures. The substrate temperature (bottom GaAs layer in Fig. 5.2) was maintained at 200 °C.

5.2.1.2. *Characterization using X-rays*

Various X-ray techniques have been used to characterize the layer growth. The most common one is RHEED. In addition, XAS, XPS, and *ex situ* high-resolution transmission electron microscopy (HRTEM) have also been used.

RHEED for CrAs epilayers The RHEED patterns of ZB CrAs were used to characterize the quality of films by Akinaga and Mizuguchi (2004). In Fig. 5.3, the RHEED patterns for a CrAs surface on GaAs along the [110] and [100] directions of the substrate are shown. The streaky features and symmetry of the intensities are the signature of the ZB structure. By looking at the temporal change of the patterns from streaky to spotty, these authors identify a critical thickness of 3 nm for the film to exhibit the ZB structure.

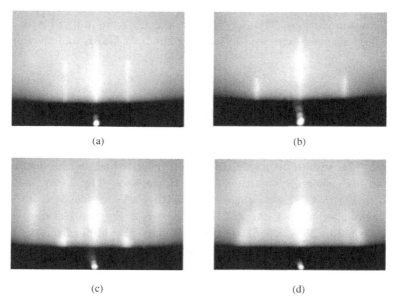

(a) (b)

(c) (d)

Fig. 5.3.   The RHEED patterns of CrAs films with the electron beam incident along the [110] (a and c) and [100] (b and d) directions of a GaAs substrate (Akinaga and Mizuguchi, 2004).

Further confirmation of the ZB structure for a film of 2 nm was obtained by HRTEM.

Etgens *et al.* (2004) also used *in situ* RHEED patterns to characterize a CrAs film. The patterns started with a well-ordered GaAs streaky form but faded into a strong diffuse background at 1 Å deposition. After 2 Å deposition, the faded streaky patterns changed to elongated spots. The spotty patterns persisted with increasing growth thickness. These authors tried to anneal the films at 300 °C under As pressure. The streak patterns became more elongated in the diffuse background. Inter-diffusion between layers did not appear as confirmed by core-level photoemission experiments.

GIXD for CrAs epilayers The grazing incidence X-ray diffraction (GIXD) technique is effective to investigate the crystallography of thin films and epilayers (Robinson, 1991) because the X-ray probes the substrate at total reflection condition to enhance the signal from epilayers with respect to the signal from the bulk. Etgens *et al.* (2004) used this scheme to examine three CrAs film samples. The thicknesses of the samples were 12, 25, and 45 Å, respectively. These samples are labeled as X12, X25, and X45. Before carrying out the GIXD experiments, the surface of each sample was cleaned and verified by Auger spectroscopy indicating that there were no C and O contaminations. The X-ray measurements were indexed with cubic GaAs(001) surfaces expressed in terms of the bulk GaAs, and they are:

$$\mathbf{a}_{GaAs}^{S} = \frac{1}{2}[1\bar{1}0]_{GaAs}, \quad \mathbf{b}_{GaAs}^{S} = \frac{1}{2}[110]_{GaAs}, \quad \mathbf{c}_{GaAs}^{S} = \frac{1}{2}[001]_{GaAs}.$$

$$(5.1)$$

The calculated CrAs lattice constant of 5.8 Å (Galanakis, 2002a) was used. This value gives a 2.6% mismatch with that of GaAs. Therefore, they searched the strained and relaxed CrAs layered samples using the GIXD method. The results displayed diffraction peaks which can be associated with an orthorhombic structure for all thicknesses. The X-ray spots are shown in Fig. 5.4. The ones associated with GaAs are denoted by $*$ and those associated with CrAs are indicated by $\diamond$. For the thicknesses less than or equal to 25 Å, the patterns are similar. They differ from the thickest case. The thinnest case has [100]CrAs parallel to [110]GaAs (the a-axis) and [010]CrAs parallel to [1$\bar{1}$0]GaAs (the b-axis). The c-axis of CrAs is parallel to the growth direction. Along the a- and b-axes, there is a 6% and 13% mismatch with respect to the corresponding lattice constants of the substrate. With this thin layer, only one epitaxy was observed. The result suggests that $\mathbf{a}_{GaAs}^{S}$ and $\mathbf{b}_{GaAs}^{S}$ in Eq. (5.1) are not equivalent.

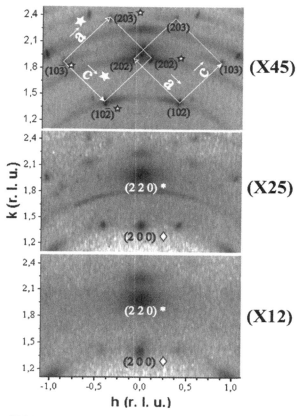

Fig. 5.4. The GIXD patterns for the three samples with different thicknesses (Etgens *et al.*, 2004). The axes are in reciprocal lattice units.

For the thickest case X45, there are two different domains. The orientations of the epilayers were determined and are denoted as (a,c) and (a^*,c^*) in Fig. 5.4. The orientations of the axes of the surface cells have been identified to be [100]CrAs ∥ [100]GaAs and [001]CrAs ∥ [010]GaAs for (a,c) and [001]CrAs ∥ [100]GaAs and [100]CrAs ∥ [010]GaAs for (a^*,c^*). The b-axis of CrAs $\mathbf{b}_{CrAs}$ is in the growth direction. From the spots, the length of a differs from $|\mathbf{a}_{CrAs}|$ by 0.23%. On the other hand, the one in the direction of c has 9.5% mismatch. These basis vectors correspond to the fully relaxed epilayer. Comparing the different GIXD patterns of the thin to thick epilayers, the single phase can be due to a metastable phase formed at the beginning of the growth. Consequently, the growth results can depend on the film thickness. Since characteristics of the ZB structure were not observed in these experiments, Etgens *et al.* (2004) attributed the failure of finding the ZB structure to the lattice mismatching. However, there is the possibility that this growth did not control the orientation of CrAs and GaAs, so that CrAs is not in the ZB structure.

RHEED and HRTEM for CrSb epilayers   The RHEED and HRTEM methods were used to characterize the substrates and film forms of CrSb (Zhao *et al.*, 2001). The HRTEM spectra exhibit features of a film in real space.

The RHEED picture shows a streaky (1×3) pattern during the CrSb growth on a Sb-terminated GaSb surface at the substrate temperature ranging from 250 °C to 400 °C. This initial pattern became weaker and disappeared after 40 seconds. For the growth on other substrates, such as $Al_{0.84}Ga_{0.16}Sb$ and GaAs, the streaky characteristic of the ZB GaAs buffer layer disappeared after depositing CrSb for 60 seconds.

The results of HRTEM for a CrSb layer grown on GaAs for 30 s at the substrate temperature of 250 °C and with the Cr flux of $2 \times 10^{13} \, cm^{-2} \, s^{-1}$ were obtained by Zhao *et al.* (2001). The film sample was capped by GaAs. Both the limited and wider areas show 1 ML of CrSb without any defects, such as dislocations, at the interfaces. The results provide evidence of a more or less ideal 1 ML growth. This characterization should serve as a standard for a δ-layer growth.

RHEED, XAS, and TEM for CrAs/GaAs multilayers   The RHEED, XAS (Mizuguchi *et al.*, 2002), and cross-sectional transmission electron microscopy (TEM) (Akinaga and Mizuguchi, 2004) have been used as the primary methods to characterize the CrAs/GaAs multilayers.

The surface conditions were monitored by RHEED during all growth processes by Mizuguchi *et al.* and Akinaga and Mizuguchi. The RHEED patterns for different samples with and without capping of GaAs layers are summarized in Table 5.2. In the table, A, B, C, and D represent different CrAs/GaAs multilayer configurations. E and F are the cases of substituting CrAs layers by alloys of (Ga,Cr)As. E is for 50% of Cr while F is for 1% of Cr. Their multilayer configurations are the same as for the multilayer A. The streaky pattern of RHEED is a characteristic of the ZB CrAs layer. In the multilayer A, the streaks related to CrAs are weaker than those of the GaAs buffer layer. The streaky pattern persists up to the tenth CrAs layer in the multilayer A. These results suggest that two layers of CrAs grown on

Table 5.2. The CrAs/GaAs multilayer samples and characteristics of RHEED analysis. x represents the number of CrAs multilayers, y is the number of GaAs multilayers, and z denotes the number of periods of these sublayers in the growth direction (Mizuguchi *et al.*, 2002).

Multilayer sample	x (ML)	y (ML)	z (period)	RHEED pattern
A	2	2	10	Streaky
B	2	1	10	Spotty
C	3	3	10	Spotty
D	2	2	100	Spotty
E				Spotty
F				Streaky

two layers of GaAs can be formed in multilayers. The spotty patterns in the multilayer B and C indicate that the growth mode is not epitaxial, where the samples have more than two layers of CrAs ($x > 2$) with less than two layers of GaAs ($y < 2$). Therefore, it is concluded that the set of $(x, y) = (2, 2)$ is the optimum condition to realize the epitaxial ZB-CrAs/GaAs multilayer. The thick multilayer D shows initially epitaxial growth. However, a change of the patterns from streaky to spotty happens gradually. At $z \sim 50$, the patterns appear to have extra spots, indicating that some unknown phase is present. For multilayer D, the RHEED eventually shows a perfect spotty pattern. It is interpreted as the CrAs layers losing the ZB structure. Akinaga and Mizuguchi (2004) increased the substrate temperature to 300 °C to grow superlattices with two layers of CrAs and two layers of GaAs up to 100 periods without any spotty pattern appearing in the RHEED.

X-ray absorption spectroscopy measurements were performed to probe the chemical bonding of the alloys using the multilayer A as the reference (Mizuguchi *et al.*, 2002). The photon energy varies continuously from 570 to 595 eV with the incident angle set at 60° from the normal. The fluorescence yield from the Cr L-line edge was measured using a Si detector with the detection angle normal to the sample surface. The results are shown in Fig. 5.5. The spectra were normalized by the peak height. The integral backgrounds were subtracted in each case. The measured spectra originated from the Cr L_3 and L_2 peaks of multilayer A. The spectra for the multilayer E are very similar. For the multilayer F, there is at least a 0.5 eV down shift with respect to the multilayer A. The possible reasons are as follows:

- In the multilayer E, CrAs can be in a segregated state while in multilayer F, the Cr atoms are in the dilute magnetic semiconductor state.

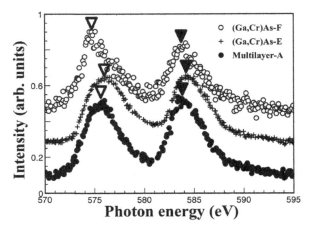

Fig. 5.5. XAS results of multilayer CrAs/GaAs samples of A, E, and F listed in Table 5.2. Unfilled and filled ▽'s denote peak positions (Mizuguchi *et al.*, 2002).

- In the multilayer E, CrAs precipitates.
- There is a possibility for Cr to diffuse into the GaAs region in the multilayer E.

The TEM image for a multilayer CrAs/GaAs with x, y, and z of 4, 4, 10, respectively, was obtained by Akinaga and Mizuguchi (2004), which clearly shows the multilayer region. The sample was grown on a 20 nm GaAs buffer layer. At the interface, there are possible defects. But there is no Cr inter-diffusion into GaAs buffer and cap layers as studied by the secondary ion mass spectroscopy (SIMS).

5.2.1.3. *Magnetic properties*

In addition to investigating the sample growths and characterizations, they also studied the magnetic properties. Among them, two groups carried out the measurements using the SQUID magnetometer.

CrAs epilayers Akinaga *et al.* (2000a) measured the hysteresis loop for an epitaxial film of CrAs grown on GaAs(001) at RT. The magnetic field was applied parallel to the film plane. The saturation magnetization is 560 emu/cm^3. This is equivalent to $3.0\,\mu_B$ per formula unit and agrees with theoretical predictions (Akinaga *et al.*, 2000a; Galanakis, 2002a; Pask *et al.*, 2003). For different samples under identical growth conditions, there is approximately 10% spread in values of the saturation magnetization. The

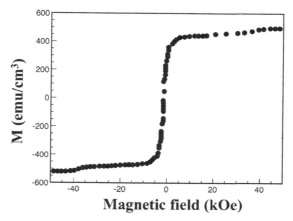

Fig. 5.6. The hysteresis loop of an epitaxial CrAs grown on GaAs. The measurement was done at 300 K (Akinaga *et al.*, 2000a).

coercivity is small. Based on these RT measurements, T_C is estimated to be greater than 400 K.

The magnetization as a function of field shown in Fig. 5.6 is a characteristic of CrAs with the ZB structure. One of the possible structures for CrAs is the MnP structure. It is known to show a helimagnetic paramagnetic transition at 265 K (Suzuki and Ido, 1993). Contamination of the Mn atoms was excluded by SIMS because the concentration of the Mn atoms is 4×10^{17} atoms/cm^3. Under this concentration, it is difficult to form FM MnAs. The ground state structure of CrAs is the NiAs structure. It is AFM with a Néel temperature of 710 K. It is unlikely that the NiAs structure contributes to the hysteresis loop at RT.

Etgens *et al.* (2004) also reported results of experiments on CrAs but did not discuss the methods employed. The main result is the saturation magnetization as a function of thickness of the films shown in Fig. 5.7. When the layers are ~4.0 nm, the saturation magnetization of different samples with the same thickness exhibits more than a factor of 3 fluctuation with significant uncertainties. At 20 Å (2 nm), the saturation value is 1000 emu/cm^3, which gives ~3 μ_B per Cr atom for an orthorhombic CrAs, assuming that only Cr atoms contribute to the magnetic moment. This number agrees with that obtained by Akinaga *et al.* (2000a). As the layer thickness increases, the magnetization reaches a value below 250 emu/cm^3. The hysteresis loops for the two films with different thicknesses are shown in the inset. The coercive field and remanent magnetization are also comparable to those given by Akinaga *et al.*

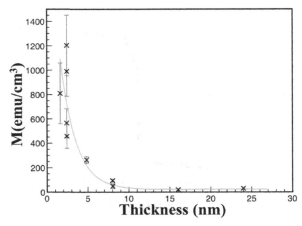

Fig. 5.7.　Magnetization as a function of thickness of a CrAs film (Etgens *et al.*, 2004).

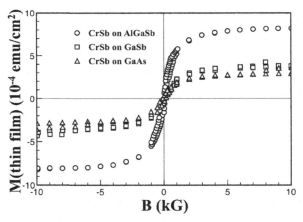

Fig. 5.8.　Hysteresis loops of CrSb thin-film samples grown on three different buffer layers – GaAs (triangles), AlGaSb (circles), and GaSb (squares) (Zhao *et al.*, 2001).

CrSb epilayers　Zhao *et al.* (2001) used SQUID to measure the hysteresis loops of CrSb thin-film samples grown on three different buffer layers. They also reported the remanent magnetization as a function of temperature. The hysteresis loops at 300 K for the CrSb thin-film samples are shown in Fig. 5.8. The contribution from the diamagnetic GaAs was subtracted. Even the layer thickness was only 1 ML; the films exhibit FM properties at RT. The saturation magnetizations at 5 K are between 3 and 5 μ_B/formula unit.

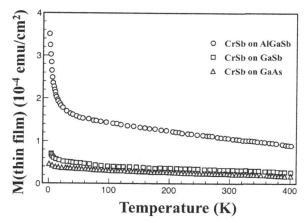

Fig. 5.9. Temperature dependence of the remanent magnetization of the three CrSb thin-film samples (Zhao *et al.*, 2001).

The value of $3.0\,\mu_B$/formula unit agrees with the theoretical calculations by Pask *et al.* (2003).

The temperature dependence of the remanent magnetization of the three samples measured at $B = 0$ is shown in Fig. 5.9. The results of the films grown on GaAs and GaSb are very close while the one grown on AlGaSb alloy is higher. At 400 K, all the samples still exhibit finite remanent magnetization indicating that the T_C is higher than 400 K.

CrAs/GaAs multilayers The SQUID measurements were used to determine the magnetic hysteresis loops of CrAs/GaAs multilayers. An example of the hysteresis loop for $[(CrAs)_2/(GaAs)_2]_{100}$ capped by 10 nm GaAs grown on a 20 nm GaAs buffer layer is shown in Fig. 5.10. The loop is very narrow. The remanent magnetization is similar to the one of the CrAa epilayer. The saturation magnetization is $400\,\text{emu/cm}^3$ and is equivalent to $2.0\,\mu_B$/formula unit, which is smaller than the theoretical value of $3.0\,\mu_B$/formula unit for bulk CrAs (Pask *et al.*, 2003).

5.2.2. *Theory*

Since the pioneering work of Akinaga and Mizuguchi (2004) on CrAs, much theoretical work has been devoted to the prediction of new HMs with ZB structure. Among them, Fong *et al.* (2004) and Fong and Qian (2004) designed superlattices with HM properties. Later, Qian *et al.* (2004a) reported an integer magnetic moment in a quantum dot composed of MnAs

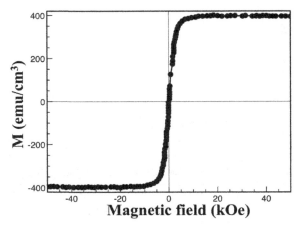

Fig. 5.10.   The hysteresis loop of a multilayer structure: $[(CrAs)_2/(GaAs)_2]_{100}$ (Akinaga and Mizuguchi, 2004).

(Qian *et al.*, 2006b). Dag *et al.* (2005) and Durgun *et al.* (2006) designed quantum wires showing half-metallic properties. Sanvito and Hill (2001) examined the half-metallic properties in δ-layer-doped Mn in GaAs. Such structures are now known as DFH.

In this section, we shall review these efforts by starting with compounds. We shall discuss the basic interactions causing the half-metallicity in these compounds. Then we shall discuss the associated quantum structures.

Despite the fact that the ZB structure is not the energetic ground state structure for the TM compounds, many half-metallic pnictides, a carbide, and several chalcogenides have been examined theoretically. Table 5.3 lists some of these theoretical efforts, including the bulk lattice constants, lattice constants at which the compound exhibits half-metallicity, and corresponding magnetic moments.

The agreement among the theories is very good for all properties except the magnetic moment; some show non-integer values. The discrepancies are mainly caused by different methods. There was some doubt whether the pseudopotential method would be suitable to treat these compounds. As illustrated in Table 5.3, however, the results obtained by the plane wave pseudopotential (PWPP) method agree very well with those of the all-electron LAPW method.

The energetic ground state structures of these compounds have also been considered. Pask *et al.* (2003) compared NiAs and ZB structures using

Table 5.3. Theoretical predictions of HMs with ZB structure.

Compound	Lattice constants (Å)		Magnetic moment (μ_B/ formula unit)	Reference
	Bulk	HM		
CrAs	5.66	5.66	3	Akinaga *et al.* (2000a) [LAPW]
	5.659	5.53–5.65	3	Galanakis (2002a) [KKR]
	5.67	5.67	3	Pask *et al.* (2003) [LAPW]
	5.67	5.50–5.87	3	Pask *et al.* (2003) [PWPP]
		5.65	3	Galanakis (2002a) [KKR]
CrP	5.35	5.48	3	Pask *et al.* (2003) [LAPW]
	5.42	5.47–5.60	3	Pask *et al.* (2003) [PWPP]
CrSb	6.15	6.15	3	Pask *et al.* (2003) [LAPW]
	6.14	5.89–6.39	3	Pask *et al.* (2003) [PWPP]
	6.139	5.69–7.17	3	Liu (2003) [LAPW]
CrSe		5.62–5.82	4	Şaşıoğlu *et al.* (2005a) [KKR]
	5.833	5.833	4	Xie *et al.* (2003) [LAPW]
MnAs	5.72	5.7	4	Pask *et al.* (2003) [LAPW]
	5.74	5.77–5.97	4	Pask *et al.* (2003) [PWPP]
		5.643	3.75	Continenza *et al.* (2001) [LAPW]
	5.70	5.70	3.5	Sanvito and Hill (2001) [Tight-binding]
		5.73–5.87	4	Şaşıoğlu *et al.* (2005a) [KKR]
MnBi	6.399	5.399	4	Xu *et al.* (2002) [LAPW]
MnGe		5.61–5.65	3	Şaşıoğlu *et al.* (2005a) [KKR]
MnP		5.308	2.73	Continenza *et al.* (2001) [LAPW]
MnSb		6.166	3.77	Continenza *et al.* (2001) [LAPW]
	6.19	6.19	4	Pask *et al.* (2003) [LAPW]
	6.21	5.96–6.46	4	Pask *et al.* (2003) [PWPP]
		6.128	4	Xu *et al.* (2002) [LAPW]
MnSi		5.52–5.65	3	Şaşıoğlu *et al.* (2005a) [KKR]
MnC		4.20–4.23	1	Şaşıoğlu *et al.* (2005a) [KKR]
	4.36	4.26	1	Pask *et al.* (2003) [LAPW]
	4.39	3.60–4.28	1	Pask *et al.* (2003) [PWPP]
VAS	5.54		1.939	Galanakis and Mavropoulos (2003) [KKR]

both pseudopotential and LAPW methods with GGA exchange-correlation. The NiAs structure is confirmed to be the ground state for the pnictides. Realizing that the ZB structure is a metastable structure and thin-film forms can be grown, a list of possible substrates (Table 5.4) for each compound was suggested. Galanakis and Mavropoulos (2003) also suggested GaAs as a substrate to grow CrAs.

In the following, we focus first on a typical pnictide, MnAs, where the underlying interactions associated with the half-metallicity will be discussed. We then discuss a TM carbide, MnC, which shows some interesting and opposite properties from the pnictides and chalcogenides.

Table 5.4.   Compound, range of lattice constants a within $\sim$4% of the corresponding equilibrium values over which half-metallicity is predicted and the suggested substrates for growth. Substrate lattice constants (Wyckoff, 1963) are given in parentheses.

Compound	Lattice constants (Å)		
	LAPW	PWPP	Substrate
CrP	>5.48	>5.47	AlP (5.53)
CrAs	>5.51	>5.50	GaAs (5.64), AlAs (5.62)
CrSb	>5.87		AlSb (6.13), GaSb (6.12), InAs (6.04)
MnAs	>5.75	>5.77	InP (5.87)
MnSb	>6.06	>5.96	AlSb (6.13), GaSb (6.12), InSb (6.48)
MnC	>4.26	>4.20	Diamond (3.57), cubic SiC (4.35)

5.2.2.1. MnAs

Several groups (Continenza *et al.*, 2001; Pask *et al.*, 2003; Şaşıoğlu *et al.*, 2005a) predicted MnAs in the ZB structure to exhibit half-metallic properties. The ZB structure of MnAs is shown in Fig. 1.9, where the filled gray circles are As atoms, and they are located at the corners and centers of the cubic faces. Open circles are the Mn atoms. It is important to note that the Mn and As atoms can form chains, for example, along the [110] direction.

With the structure in mind, we now discuss the half-metallic and electronic properties.

Density of states   The DOS is a convenient physical quantity to exhibit half-metallic properties of a compound. Another quantity is the magnetic moment/unit cell (or formula unit). In Fig 5.11, the DOS of MnAs at the lattice constant of 5.77 Å is shown (Pask *et al.*, 2003). The upper panel shows the DOS of the majority- ($\uparrow$) spin channel and the lower panel is for the minority- ($\downarrow$) spin states. E_F is set to be zero on the energy scale. The features of the DOS are:

• At E_F, a finite DOS, 0.77 states/eV-cell, appears in the $\uparrow$ channel.
• For the $\downarrow$ spin states, E_F falls within a gap of 1.70 eV.
• P value at E_F is 100%.

The position of E_F in the insulating channel is lattice-constant-dependent. When the lattice constant increases, E_F drops toward the top of the valence band. As a result, the half-metallicity disappears when the lattice constant reaches a critical value.

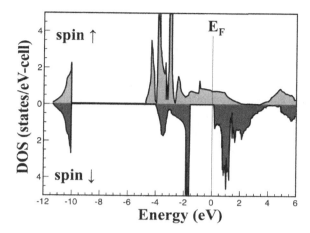

Fig. 5.11.   The density of states of MnAs (Pask *et al.*, 2003).

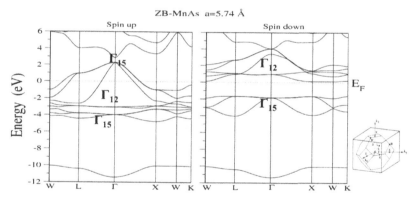

Fig. 5.12.   Band structure of MnAs along symmetry directions of the BZ (Pask *et al.*, 2003). The BZ of the fcc lattice is shown on the right.

Magnetic moment/unit cell   The calculated magnetic moment for MnAs is $4.0 \mu_B$/unit cell. Because the total number of electrons in a unit cell is an integer, the insulating channel requires integer number of electrons to fill the valence bands, and the *g*-factor of an electron is 2, the magnetic moment should then be an integer. This is a necessary condition for a compound to be an HM.

Electronic band structure   The spin-polarized band structure of MnAs from the pseudopotential method (Pask *et al.*, 2003) is shown in Fig. 5.12

along a few symmetry directions of the BZ for the fcc structure. As a reference, the BZ is shown in Fig. 5.12. The lowest bands in both spin channels are formed by the s-like state of the As atom. The next set of bands at the Γ-point (labeled as Γ_{15}) are triply degenerate t_{2g} states. They are the bonding states as the consequence of d-p hybridization. They have predominantly the character of the As p-like states. The next higher energy bands are denoted as Γ_{12} or e_g. They are doubly degenerate and are called the nonbonding d-states of the Mn atom. The lobes of the charge distribution of these two states point toward the second neighbors of the Mn atoms. In this cubic environment, it is possible to identify that they are originated from the d_{z^2} and $d_{x^2-y^2}$ states. In the $\uparrow$ spin channel, these doubly degenerate states are occupied, while those in the $\downarrow$ spin channel lie above E_F. These bands separate from the lower energy Γ_{15} (t_{2g}) bands by a gap. Therefore, the minority channel exhibits insulating behavior. In the majority-spin ($\uparrow$) states, we label another Γ_{15} (t_{2g}^*) above Γ_{12}. These are antibonding states resulting from d-p hybridization. The broad features of these bands make them partially occupied along the Γ–X direction contributing to the metallic behavior of the majority-spin states. These features show the half-metallic properties from the band structure point of view.

Charge densities   To examine the bonding properties in this ZB half-metallic compound, the charge densities of the occupied t_{2g} and e_g states associated with the majority-spin channel are shown in Fig. 5.13. Both

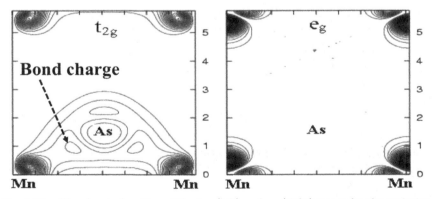

Fig. 5.13.   The charge densities of the t_{2g} (left) and e_g (right) states for the majority-spin states. The horizontal axis is in the [110] direction and the vertical axis is the z-axis of a cube (Pask *et al.*, 2003).

sections contain the zig-zag chain formed by the Mn and As atoms along the [110] direction.

The left panel shows the charge distribution of the t_{2g} states. The bond charge is clearly exhibited between the two atoms. In the present case, the high concentration at the Mn atom manifests the tightly bound character of the d-state. If we translate this unit cell in both x and y directions, the four lobes of the d-state can be identified. It is a linear combination of the d_{xz} and d_{yz} states. The right panel is for the e_g states. We can easily identify the lobes around the Mn atom associated with the d_{z^2} state. These lobes point toward the second neighbor Mn atoms. There is no distribution pointing toward the nearest neighbor As atoms. The nonbonding characteristic of these states is clearly demonstrated.

5.2.2.2. *MnC*

Half-metallic phases Among all the compounds studied by Pask *et al.* (2003), MnC shows interesting half-metallic properties. The majority- and minority-spin channels reverse their roles for the metallic and insulating behaviors as compared to the pnictides and chalcogenides. The calculated magnetic moment is only $1.0 \, \mu_B$/unit cell which will be explained in Section 5.2.2.3. Qian *et al.* (2004b) found that MnC in fact has two different ranges of lattice constants for which the half-metallic properties exist. The lattice-constant-dependent total energy and magnetic moment are shown in Fig. 5.14. From the total energy curve (the lower panel), the optimized

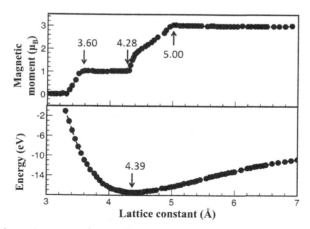

Fig. 5.14. Magnetic moment/unit cell and total energy as a function of lattice constant in MnC (Qian *et al.*, 2004b).

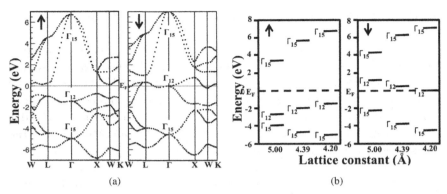

Fig. 5.15. (a) The band structure of MnC at $a = 4.20$ Å, and (b) the energy levels at the Γ-point as a function of lattice constant (Qian *et al.*, 2004b).

lattice constant is found to be 4.39 Å. The interesting behavior of the magnetic moment/unit cell is also shown. For lattice constant greater than 5.0 Å, the magnetic moment is $3.0\,\mu_B$/unit cell and can be accounted for by the ionic model. For lattice constant between 3.28 and 4.23 Å, there is another region with integer magnetic moment. The value of the magnetic moment is $1.0\,\mu_B$/unit cell.

Band structure at smaller lattice constants The band structures for lattice constants at 5.0 and 4.20 Å were calculated (Qian *et al.*, 2004b). The band structure at $a = 5.0$ Å is similar to the one of MnAs and is not given. The band structure at $a = 4.20$ Å is shown in Fig. 5.15(a). The lowest s-band is not shown. The ordering of the bonding, nonbonding, and antibonding states indicated by the energy levels at Γ (Fig. 5.15(b)) are the same as in MnAs.

Periodic local moment There is another interesting property of MnC at lattice constants between 3.60 and 4.28 Å. Under such a condition, the magnetic moment of MnC is still determined by the majority-spin states. However, the majority-spin states exhibit insulating properties. The corresponding charge density is localized in the bonding region and around the Mn atom. Therefore, the local magnetic moment can show a periodic distribution. At low temperature, these moments can have spin-wave excitations. An electron in the minority-spin channel occupying a state at E_F can scatter with the spin waves at low temperature. It is anticipated that the resistivity should show a T^5 characteristic (Qian *et al.*, 2004b).

Because of these interesting properties, it is worth exploring the possibility of growing this compound by choosing the proper substrates. BP, BN, and SiC have been suggested (Fong and Qian, 2004; Fong *et al.*, 2008). Based on the results predicted by Fong and Qian, only a monolayer MnC could be grown on a SiC(100) substrate.

Underlying interactions for half-metallicity in ZB structure

There are three types of interactions involved in the half-metallicity in TM compounds with ZB structure. The first is due to the environment of the atoms in the unit cell — the crystal field. The second is due to the interaction between nearest neighbors — the d-p hybridization. And the third is a magnetic interaction — the exchange interaction.

Crystal field As shown in Fig. 1.9, each of the two atoms in a unit cell is surrounded by four neighbors forming a tetrahedron. Neighboring ions of the TM atom exert Coulomb fields on the d-states of the atom. The effect is to split the five-fold degeneracy of the d-states into triply degenerate t_{2g} and doubly degenerate e_g states. Therefore, a gap between the t_{2g} and e_g states is formed.

d-p hybridization The triply degenerate t_{2g} states of the TM element are composed of d_{xy}, d_{yz}, and d_{zx} states. Their linear combinations form orbitals pointing toward the nearest neighbors of the metal element. The sp^3 orbitals of the non-metal atoms point their lobes toward their nn. The orientations of a d-orbital and sp^3 orbital are shown schematically in Fig. 1.9. The d- and p-states overlap and hybridize. This hybridization forms the bonding and antibonding gap characterized by the two Γ_{15} states at the Γ-point in the band structure. These two interactions can be summarized in terms of energy levels as shown in Fig. 1.10. The Coulomb fields, due to the neighboring TM elements, cause the s- and the p-states of the As atom to form sp^3 directional orbitals. Under this circumstance, the p-states remain degenerate.

Exchange interaction Up to this point, the bands in a compound are spin degenerate. Now, the exchange interaction lifts the degeneracy of the ↑ and ↓ states and shifts the ↓ state energies upwards.

With the proper combinations of these three interactions, E_F intersects one or more of the bands of the majority-spin channel and falls in a gap of the minority-spin states. The compound is then an HM. A reversal of roles played by the majority- and minority-spin states is also possible due to variations of the three interactions, as manifested in MnC.

The strengths of the three interactions can be estimated from the band structure. Take MnAs as an example. The energy difference between Γ_{12} and the lowest energy Γ_{15} states can be used to specify the strength of the crystal field. The difference of energies for the bonding and antibonding Γ_{15} states can be used to estimate the magnitude of the hybridization. A more refined approach is to determine first the average energies for the states derived from the two Γ_{15} states and then take the difference. The energy difference of the lowest Γ_{15} states for the majority- and minority-spin channels is related to the strength of the exchange interaction.

To understand the smaller lattice constant of MnC in terms of the strength of the three interactions, we compare the band structures shown in Fig. 5.12 and Fig. 5.15(a). The major differences shown in Fig. 5.15(a) are: (i) E_F intersects the minority-spin states; (ii) the gap is formed within the majority-spin states. The lattice-dependent energy levels at the Γ-point are shown in Fig. 5.15(b). For the majority-spin channel, the bonding–nonbonding and bonding–antibonding gaps increase as the lattice constant decreases. These results indicate that the strength of the crystal field and p-d hybridization increase under compression. The minority-spin states show similar behavior, except that the nonbonding (Γ_{12}) state is lowered with compression. This may be due to the fact that the Γ_{12} states are not occupied. The increase of dispersion of the bonding and antibonding states with respect to the case of larger lattice constant causes E_F to shift. The strength of the exchange interaction, gauged by the differences of the lower Γ_{15} states of both $\uparrow$ and $\downarrow$ states, is lessened as the lattice constant decreases.

5.2.2.3. *Qualitative explanation of magnetic moments*

The magnetic moment of MnAs is $4.0\,\mu_B$/unit cell which can be qualitatively understood in terms of the so-called "ionic model" (Schwarz, 1986) based on:

Atomic configurations   The atomic configurations of the valence states of the Mn and As atoms are $(3d)^5(4s)^2$ and $(4s)^2(4p)^3$, respectively.

Charge transfer   Since As is a group-V element, it has a large ionicity. Three electrons from the Mn atom are transferred to the As atom to form bonding and antibonding states through d-p hybridization. These three electrons pair their spins with p states from the As atom and therefore do not contribute to the magnetic moment of the crystal. Four d-electrons remain at the Mn site.

Hund's first rule Since d-states can accommodate five electrons in each spin channel, the four electrons can align their spins as required by Hund's first rule. This is essentially the manifestation of Pauli's exclusion principle and the minimization of the Coulomb interaction considering the position and spin as two dynamical variables to specify the states of the electrons. Since the four d-electrons have the same spin states, they cannot occupy the same spatial position. They are kept apart from one another. Consequently, the effect is to reduce the Coulomb repulsion. The compound can therefore have lower energy. The total spin moment of the compound is $2\hbar$. With the g-factor of 2 for each electron, the resultant magnetic moment is $4.0\,\mu_B$/unit cell. The contribution from the spin-orbit interaction to the magnetic moment is negligible due to electrons occupying 3d states.

It should be noted that for an HM the magnetic moment/unit cell should be an integer. The reasons are as follows:

- The total number of electrons is an integer in a compound.
- One of the spin channels, say the $\downarrow$ channel, is insulating. Therefore, it is necessary to have an integer number of electrons, $N_\downarrow$, to fill up the top of the valence band.
- The number of electrons in the conducting channel, $N_\uparrow$, must then be an integer.

The magnetic moment/unit cell is:

$$M = N_\uparrow - N_\downarrow. \tag{5.2}$$

Therefore, for any theoretical prediction of a compound to be an HM, it is necessary to have an integer value of the magnetic moment/unit cell.

The above explanation applies to MnC having a large lattice constant. Can we understand why the spin moment changes from 3.0 to 1.0 μ_B/unit cell in MnC as its lattice constant decreases? Let us start with the case having a larger lattice constant, 5.0 Å. From the ionic model, the three electrons remaining at the Mn atom align their spins. Two of them occupy the Γ_{12} related states and one occupies the low-energy tail of the anti-bonding state (see Fig. 5.15 for reference). At this lattice constant, there is ample space for the electrons to spatially avoid each other consistent with Pauli's principle. As the lattice constant decreases, the charges of the d-states around the Mn atom are pushed closer to the C atom as manifested by the increase of the bonding (lower Γ_{15} states) and antibonding (upper Γ_{15} states) gap and the crystal field effect (Fig. 5.15(b)). As the lattice constant decreases further, we take the value of 4.20 Å as an example, the space for the three d-electrons to align their spins becomes so limited that one of the three majority-spin electrons flips its spin. This effect is demonstrated in Fig. 5.15(b) by the lowering in energy of the Γ_{12} states in the minority-spin channel to accommodate this spin-flipped electron. To be more explicit, the magnetic charge (spin density) distribution, defined as the difference between the ↑ spin and ↓ spin charge densities, of the Γ_{12} related states in the minority-spin channel is plotted in Fig. 5.16. The section of the charge distribution is the (110) plane including the zig-zag chain of Mn and C atoms. The dot-dash contour denotes the zero magnetic charge density. The solid and the dotted contours exhibit the positive and negative values of the density.

A possible method of detecting volume-dependent half-metallic phases is to carry out the measurement of the saturation magnetic moment of the sample. If the saturation magnetization is over $500\,\mathrm{emu/cm^3}$, then the sample has $3\,\mu_B$/unit cell. If the measured result is around $180\,\mathrm{emu/cm^3}$, the corresponding saturation magnetization is $1\,\mu_B$/unit cell.

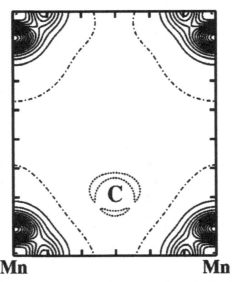

Fig. 5.16. Magnetic charge densities of MnC at 4.20 Å in the (110) plane including the zig-zag chain of the Mn and C atoms.

5.3. Half-metallic Superlattices

In addition to CrAs/GaAs multilayered structures grown by Akinaga and Mizuguchi (2004), there are two types of superlattices studied theoretically. One is the superlattice composed of half-metallic TM pnictides with the ZB structure (Fong *et al.*, 2004). The other one is a combination of one half-metallic pnictide with a semiconductor (Fong *et al.*, 2008). We shall discuss them separately.

5.3.1. *CrAs/MnAs superlattices*

The first theoretical report on the $(CrAs)_n/(MnAs)_n$ superlattices was by Fong *et al.* (2004) with $n = 1$ and 2. Since the qualitative features for these two cases are very similar, the following discussions will be focused on $n = 1$. In Fig. 5.17, the supercell for $(CrAs)_1/(MnAs)_1$ is shown. The three axes of the supercell are defined as: $\mathbf{a}_1 = (a/\sqrt{2}, a/\sqrt{2}, 0)$, $\mathbf{a}_2 = (-a/\sqrt{2}, a/\sqrt{2}, 0)$, and $\mathbf{a}_3 = (0, 0, a)$, where a is the cubic edge of the conventional cubic cell. The Cr atoms are shown by the gray circles, the Mn atoms are denoted by the filled circles and the As atoms are depicted by the open circles. The value of a in Fig. 5.17 is optimized to be 5.70 Å and is in between the lattice constant of CrAs (5.66 Å) and MnAs (5.77 Å).

Total and partial density of states   The total and partial DOS (PDOS) of the $(CrAs)_1/(MnAs)_1$ superlattice are shown in Fig. 5.18. The As occupied s-states form bands below 10 eV from E_F which is set to be zero. For the occupied d-states, the Mn d-manifold is located about 2 eV lower than

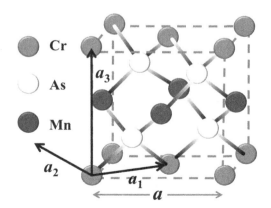

Fig. 5.17.   The supercell model of $(CrAs)_1/(MnAs)_1$.

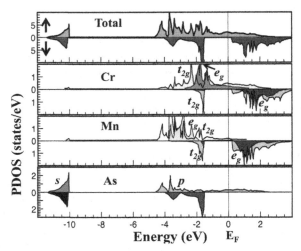

Fig. 5.18.   Total and partial DOS of the majority- and minority-spin channels. TDOS
are in units of states/eV-unit cell.

those of the Cr d-states. This is expected because Mn has valence 7 while
Cr has 6. The e_g states of both atoms are higher than the center of gravity
of the t_{2g} states. As shown in Fig. 1.10, under the crystal field the t_{2g} states
have higher energy than the e_g states. The t_{2g} states in the superlattice are
the hybridized bonding states. Another feature shown in the PDOS is the e_g
states of both metallic elements in the minority-spin channel located above
E_F. The valence states of this channel are predominantly the As p- and
Mn t_{2g}-hybridized states.

Charge densities   Total charge densities in a section containing the
zig-zag chain of the two TM elements and As atom are shown in Fig. 5.19
for the two spin channels. The contours reflect the features of the DOS and
PDOS. In the majority-spin channel, the bond charges are explicitly indicated by contours located between the Mn and As atoms. For the minority-
spin states, contours around the two TM elements exhibit the four-lobe
characteristic of the d-states.

Magnetic moment and magnetization   One of the key quantities for
spintronic applications is the magnetic moment. Based on the assumption
that the exchange splitting is sufficient to push the minority e_g states above
E_F, with minority anion-s and p-t_{2g} hybrid states fully occupied, Pask *et al.*
(2003) suggested the following relation for the magnetic moment/unit cell

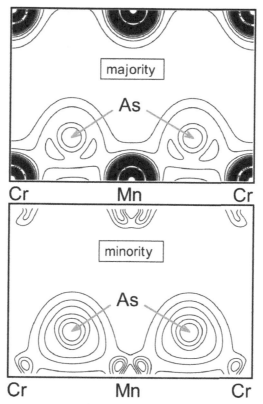

Fig. 5.19. Total valence charge densities of $(CrAs)_1/(MnAs)_1$ of majority-spin states (upper panel) and minority-spin channel (lower panel) in a section containing the zig-zag chain of the two TM elements and As atom.

(M) of an HM in the ZB structure:

$$M = N_{majority} - N_{minority} = Z_{total} - 2N_{minority}, \qquad (5.3)$$

where Z_{total} is the total number of valence electrons, and $N_{majority}$ and $N_{minority}$ are the number of the majority- and minority-spin states, respectively. For $(CrAs)_1/(MnAs)_1$, $Z_{total} = 23$ and $N_{minority} = 8$ give $M = 7\mu_B$. Similarly, Eq. (5.3) gives a moment of $14\mu_B$ for $(CrAs)_2/(MnAs)_2$. These values are confirmed by numerical calculations. They are the sum of M for each constituent compound making up the superlattice. It is important to note that the increased magnetic moment/unit cell for larger cells implied by Eq. (5.3) does not necessarily mean an increase

Table 5.5. Total DOS at E_F of the majority-spin channel, E_g of the minority-spin states, the magnetic moment/unit cell for $(CrAs)_1/ (MnAs)_1$, and $(CrAs)_2/(MnAs)_2$. Superscript a refers to Pask *et al.* (2003).

Sample	DOS at E_F (states/eV-cell)	E_g (eV)	M (μ_B)
$(CrAs)_1(MnAs)_1$	1.94	1.65	7.0
$(CrAs)_2(MnAs)_2$	3.47	1.62	14.0
$CrAs^a$	0.85	1.85	3.0
$MnAs^a$	0.77	1.70	4.0

of the saturation magnetization. The latter is defined as the magnetic moment density. Both superlattices have the same saturation magnetization of $672.8\,emu/cm^3$. Comparing the saturation magnetizations of CrAs ($572.4\,emu/cm^3$, $3.0\,\mu_B$) — which agrees well with the measured $560\,emu/cm^3$ — and MnAs ($763.2\,emu/cm^3$, $4.0\,\mu_B$), the calculated value for the superlattices is approximately the average of the two constituents. Therefore, it is not viable to increase the saturation magnetization by growing superlattices. These results are summarized in Table 5.5 along with DOS at E_F for the majority-spin channel and energy gap (E_g) of the minority-spin states.

Substrates for growth To determine possible substrates for growing CrAs/MnAs superlattices, Fong *et al.* (2004) calculated the half-metallicity for a range of lattice constants. Between 5.60 and 6.03 Å, the superlattices retain their half-metallic properties. This range spans the experimental lattice constants of AlAs (5.62 Å), GaAs (5.65 Å), and InP (5.81 Å). These semiconductors can serve as substrates imposing the least strain for growing CrAs/MnAs superlattices to exhibit half-metallic properties.

5.3.2. *Superlattice showing spin-polarized ballistic transport*

Unit cell In all theoretical efforts, the objective has been whether the sample exhibits half-metallic properties. Qian *et al.* (2004b) designed a half-metallic superlattice which can exhibit ballistic transport properties. The supercell consists of one layer of GaAs, one layer of MnAs, two layers of CrAs, and two capping layers of GaAs. The supercell model is shown in Fig. 5.20. The unit cell is outlined by the solid and dashed lines. The a-axis is along the [110] direction of the conventional cell in the ZB structure. $|a| = \sqrt{2}a_o$ and $|b| = 3a_o$, where $a_o = 5.722$ Å is the optimized lattice constant for GaAs.

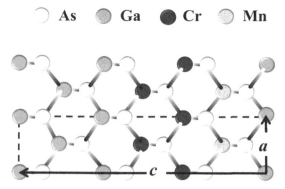

Fig. 5.20. A superlattice model exhibits the spin-polarized ballistic transport properties. a is along the [110] direction of the conventional cubic cell (Qian *et al.*, 2004b).

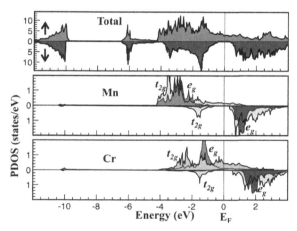

Fig. 5.21. Total DOS and PDOS of a superlattice showing the ballistic transport (Qian *et al.*, 2004b).

Total and partial density of states The total density of states (TDOS) and the PDOS are shown in Fig. 5.21. The qualitative features are similar to those of CrAs/MnAs superlattice discussed in Section 5.3.1. The majority-spin states exhibit metallic behavior (the top panel of Fig. 5.21). They are primarily contributed by the d-states of the Cr atom (the third panel). The occupied d-manifold of Cr is higher in energy than that of the Mn atom (the second panel). The e_g states of the Cr atom are occupied. The minority-spin channel exhibits a gap at E_F. The gap value is 0.95 eV and is nearly

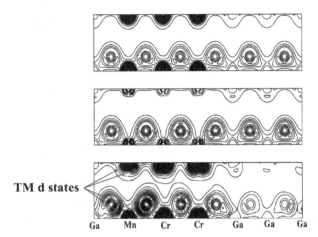

Fig. 5.22. Total charge densities of ↑ spin states (top), ↓ spin states (middle), and the total (bottom) in an energy ranging between E_F and $E_F + 0.3$ eV. The section contains the zig-zag chain of the TM elements, Ga, and As atoms (Qian *et al.*, 2004b).

half of the values of CrAs (1.88 eV) and MnAs (1.74 eV). The occupied d-states are the bonding t_{2g} states and those e_g states are above E_F.

Charge densities In Fig. 5.22, the charge densities are plotted in a section consisting of a zig-zag chain of atoms. There are three panels. The top panel shows the total valence charge density of ↑ spin states. In the region between the Mn and Cr atoms, the distribution is similar to the CrAs/MnAs superlattice. The As atoms are located between the labeled atoms as their nn to form a chain. In the GaAs regions, bond charges are formed between the Ga atoms and their nn. The middle panel shows the charge distribution of ↓ spin states. Similar to the CrAs/MnAs case, the four lobes associated with d-states of Cr and Mn atoms are illustrated. They are mainly the t_{2g} type of bonding states.

Ballistic conductance In order to see the conducting channel of this superlattice along the direction perpendicular to the layers, Qian *et al.* (2005) plotted the charge density of the states in the majority-spin channel and within 0.3 eV above E_F. This charge distribution exhibits the coherent extended feature, allowing an electron injected into the superlattice at the left side of GaAs region by an electric field whose strength does not disturb the charge distribution in the superlattice. Consequently, the electron travels through the superlattice without suffering any scattering. In addition,

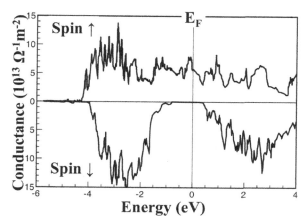

Fig. 5.23. Ballistic conductance along the c-axis for the superlattice is shown as a function of energy (Qian *et al.*, 2005). E_F is set to be zero.

they also calculated Fermi velocity in the travelling direction. This velocity relates to the ballistic conductance G_σ by the following relation:

$$G_\sigma = \frac{Ae^2}{2m} \frac{1}{8\pi^3} \sum_n \int \langle \Psi_{\mathbf{k}n,\sigma} | P_z | \Psi_{\mathbf{k}n,\sigma} \rangle \delta(\varepsilon_{\mathbf{k}n,\sigma} - E_F) d^3k, \qquad (5.4)$$

where σ labels the spin channel. A is a finite cross section, e is the electronic charge, P_z is the z-component of momentum operator. $\Psi_{\mathbf{k}n,\sigma}$ is the wave function for band n and at a $\mathbf{k}$ point inside the BZ with energy $\varepsilon_{\mathbf{k}n,\sigma}$. The results of the conductance for the two spin channels are shown in Fig. 5.23. The upper panel is for the $\uparrow$ channel and shows a finite conductance at E_F. The conductance of the $\downarrow$ states is shown in the lower panel. There is zero conductance at E_F. All these results illustrate that the superlattice shown in Fig. 5.20 can have a spin-polarized ballistic transport. These kinds of superlattices can be used to fabricate spin filters.

5.4. Quantum Dots

Advances in nanoscience are based on the studies of superlattices, quantum dots, and other quantum structures with length on the order of nanometers. Having the possibility of growing HMs in thin-film form, it is natural to explore the possibilities for spintronic applications using quantum dots. There have been several attempts (Ono *et al.*, 2002; Okabayashi *et al.*, 2004) at growing quantum dots using compounds predicted to exhibit half-metallic properties. Theoretical designs of half-metallic quantum dots have also been initiated by Qian *et al.* (2006b). We shall discuss first the experimental efforts then the theoretical results.

5.4.1. *Experiment*

Growths of quantum dots composed of TM pnictide, such as MnAs, have been reported by Ono *et al.* (2002) and Okabayashi *et al.* (2004), respectively. We discuss the growths first, then characterizations, and finally, the experimental determination of the electronic properties.

5.4.1.1. *MnAs quantum dots*

Growth   To grow MnAs quantum dots with the MBE technique, Ono *et al.* used a scheme developed by Akinaga *et al.* (2000b) for growing MnSb granular films on the sulfur-passivated GaAs substrate. The reason for passivating GaAs by a group-VI element is to lower the surface energy so that self-assembled growth can take place. Ono *et al.* passivated the n^+ GaAs(001) substrate with the S atoms by first dipping the substrate into a solution of $(NH_4)_2S_x$ for one hour then rinsing the substrate with pure water. The substrate was then heated to 200 °C. The source beams were finally switched on with the flux ratio of As/Mn set to be 4–5. To ensure that the MnAs quantum dots are in the ZB structure, these authors also grew the NiAs-type bulk MnAs and $Ga_{1-x}Mn_xAs$ films on GaAs(001) substrates for comparisons. For the $Ga_{1-x}Mn_xAs$ films, they grew first a 15 nm GaAs buffer layer on a GaAs substrate after the removal of an oxide layer at the surface by heating the substrate to 580 °C. Then the temperature of the buffer layer was cooled to 200 °C. Another GaAs buffer layer of 10 nm was grown on top. Finally, the films were grown. For the alloy and bulk films, the ratio of As/Mn is also set at 4–5.

Characterization of samples   After rinsing away $(NH_4)_2S_x$ and setting the substrate temperature at 200 °C, the RHEED pattern changed from a halo to a 1×1 streaky form. During the growth of MnAs quantum dots, the pattern became spotty.

To probe whether the MnAs quantum dots are in the ZB structure, high-resolution cross-sectional TEM and selected-area electron diffraction (SAED) were used. The results are shown in Fig. 5.24. On the left, the TEM images of two cross sections of the interfaces between the MnAs and GaAs substrate are exhibited. The details of the images are very similar. On the right, the SAED patterns are shown. The patterns for the two sections are

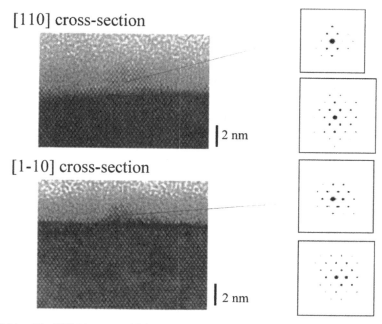

Fig. 5.24. The TEM image and SAED patterns of nanoscale MnAs and GaAs substrates (Ono *et al.*, 2002).

also very similar. These results indicate that both the quantum dots and substrate have the same structure. The lattice-constant mismatch between the MnAs and substrate can also be detected by both TEM and SAED. A 0.7% mismatch was estimated in the plane parallel to the interface.

Size distribution The size distribution of MnAs quantum dots can be determined by high-resolution scanning electron microscopy (SEM). The results obtained by Ono *et al.* (2002) are shown in Fig. 5.25. Most of the dots have diameters near 16.3 nm.

Electronic properties To probe the Mn 3d partial DOS of the nanoscale MnAs dots, Okabayashi *et al.* (2004) used the 3p–3d resonant photoemission spectroscopy. The photon energies were between 46 and 55 eV and the light was not polarized. These authors compared spectra of different density of quantum dots to those of an *in situ* prepared alloy in ZB-like structure and a NiAs-like MnAs film.

The comparisons of the on-resonance (hν = 50 eV) photoemission spectra are shown in Fig. 5.26. When the photon energy is at and above 50 eV,

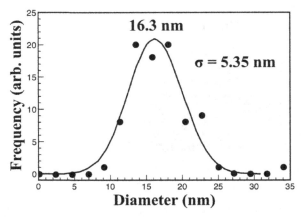

Fig. 5.25. Size distribution of MnAs quantum dots determined by SEM (Ono *et al.*, 2002).

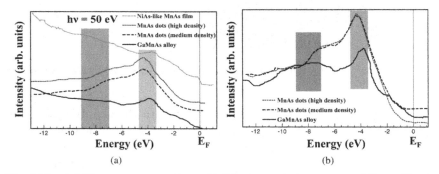

Fig. 5.26. (a) The on-resonance ($h\nu = 50\,\text{eV}$) photoemission spectra of $Ga_{1-x}Mn_xAs$ alloy, medium-density quantum dots, high-density quantum dots, and a NiAs-like MnAs film. (b) The difference spectra between the on-resonance ($h\nu = 50\,\text{eV}$) and off-resonance ($48\,\text{eV}$) spectra can be obtained as a measure of the Mn 3d partial DOS (Okabayashi *et al.*, 2004).

a peak at $4.0\,\text{eV}$ below E_F appears in all spectra. A shoulder follows in the low-energy side (around $-7 \sim -9\,\text{eV}$) of the peak only for the alloy and quantum dots but not for the film. It was therefore concluded that this is more evidence showing that the quantum dots have the ZB structure.

Magnetic properties Magnetic properties of MnAs quantum dots were measured by Ono *et al.* (2002) using SQUID. From Fig. 5.27(a), the FM ordering in the MnAs quantum dots can be clearly seen in the hysteresis

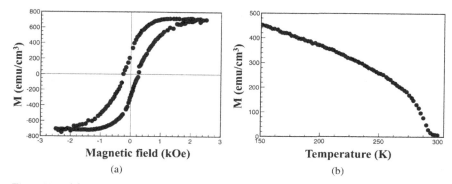

Fig. 5.27. (a) Hysteresis loop, and (b) the temperature dependence of the magnetization in MnAs quantum dots (Ono *et al.*, 2002).

loop measurement at 50 K. The T_C for the quantum dots is estimated to be 280 K based on the temperature dependence of the magnetization measurement (Fig. 5.27(b)).

NiAs-type bulk MnAs and $Ga_{1-x}Mn_xAs$ films on GaAs(001)
The bulk MnAs in the NiAs structure is hexagonal. The TEM image and SAED pattern are distinct from GaAs but similar to MnAs quantum dots grown on a GaAs(001) substrate without any passivation of the S atoms. For the alloys, the low-temperature growth on the 10 nm buffer layer shows a 1×2 surface reconstruction in the RHEED pattern. The results differ from those for the quantum dots.

5.4.2. *Theory*

5.4.2.1. *MnAs quantum dot*

Qian *et al.* (2006b) built a model of MnAs quantum dot in the ZB structure. The model, a cluster, is composed of a total of 41 atoms. They calculated the electronic properties and magnetic moment using *ab initio* pseudopotential method and DFT with GGA exchange-correlation.

Cluster model The cluster was built by choosing a center atom in the cubic crystalline MnAs and by retaining up to the third shells of the atoms. In terms of the number of atoms in each shell, the cluster is specified as $MnAs_4Mn_{12}As_{24}$. It has a radius of 6.1 Å. To saturate the dangling bonds at the edge of the cluster, sixty hydrogen atoms were used. Since these H atoms are attached to the outer shell As atoms, their effective charge is chosen at 0.75 e, where e is the electron charge. The cluster is put in a cubic supercell with a lattice constant of 22.96 Å, four times the size of the bulk MnAs with the ZB structure. The nearest-neighbor distance between atoms in neighboring clusters is 4.06 Å.

Effects of relaxation   The cluster was relaxed in both FM and AFM configurations with respect to the shells of the Mn atoms. After relaxation with the FM configuration, the average bond length of $MnAs_4$ at the center is 2.491 Å, which is larger than the unrelaxed value of 2.486 Å. When the spin of the center Mn atom is flipped (AFM configuration), the corresponding average bond length is reduced to 2.470 Å. This contraction is caused by the so-called exchange striction (Solovyev and Terakura, 2003). The total energy of the AFM configuration is lower by 125 meV with respect to the FM configuration. Based on this result, the AFM phase is energetically more stable than the FM phase.

Electronic properties and magnetic moment   The total and partial DOS for the spin channels were calculated and are shown in Fig. 5.28 for the FM phase and AFM phase, respectively. In the AFM configuration, the spin of the central Mn is oriented opposite the spins of the rest. The partial DOS are the projected DOS on one atom. As usual, E_F is set to be zero. The major difference between FM and AFM is that in FM the minority-spin channel exhibits a gap of 1.83 eV as compared to 0.97 eV in AFM. The calculated magnetic moments for FM and AFM are 52 μ_B/cluster and 42 μ_B/cluster, respectively. With the integer values of magnetic moments

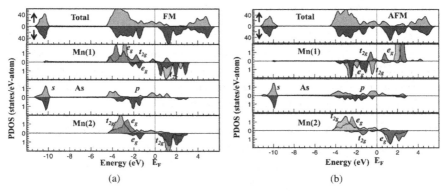

Fig. 5.28.   Total DOS (top panels) and PDOS for the FM phase (a) and AFM phase (b) of MnAs quantum dot (Qian *et al.*, 2006b).

and gaps appearing in the minority-spin channel, the cluster in either the FM or AFM phase can be an HM based on the criterion of having an integer magnetic moment of the sample.

The major contribution of d-states in both FM and AFM phases is in the region between -1.0 and $-4.5\,$eV. In the majority-spin channel, they contribute dominantly in the region between -3.2 and $-4.0\,$eV (Fig. 5.28(a) and (b)). By examining the PDOS, there appears a difference between the two magnetic phases. In Fig. 5.28(a), these states are originated from the t_{2g} states in the majority-spin channel of the Mn atom at the center of the cluster and the second-shell Mn atoms. These d-states hybridize with p-states of their neighboring As atom. The p-states are concentrated at $-4.0\,$eV. In Fig. 5.28(b), on the other hand, the t_{2g} states are primarily from the minority-spin states of the second-shell Mn atoms. These d-states hybridize with p-states of the As atoms.

Comparing to the spin-unpolarized photoemission data, the theoretical results can account for the measured peak shown in Fig. 5.26. From the PDOS in Fig. 5.28, the origin of the peak can be identified from the hybridized bonding p-states of the As atoms. Near the region of the measured peak, d-states of the Mn atom contribute as well. It will be interesting to carry out spin-polarized photoemission measurements to distinguish the FM and AFM phases and other measurements, such as PCAR and positron annihilation, to show half-metallic properties in these MnAs quantum dots. If the FM phase is the stable one and the half-metallicity can be exhibited, then these quantum dots can potentially be one of the future materials for spintronic devices.

5.5. Digital Ferromagnetic Heterostructures

A DFH is defined as an ideal δ-doping semiconductor with TM elements. It was first attempted by Kawakami *et al.* (2000) to grow such heterostructure in the form of doping GaAs(100) with Mn. It is now called (Ga,Mn)As-DFH. A theoretical study of this DFH was carried out by Sanvito and Hill (2001). The possible enhancement of FM coupling in (Ga,Mn)As-DFH has been examined by Wang and Qian (2006). (Ga,Mn)Sb-DFH has been grown by Chen *et al.* (2002). Recently, Qian *et al.* (2006a) designed a DFH involving a δ-layer doping Si with Mn atoms. It is called Si-based DFH. The interstitial sites for Mn in Si were investigated by Wu *et al.* (2007).

5.5.1. *Experiment*

In the following, the growth and characterization of (Ga,Mn)As- and (Ga,Mn)Sb-DFH will be discussed first, then focus will be on the associated physical properties.

5.5.1.1. *Growth and characterization of (Ga,Mn)As-DFH*

The MBE method was employed by Kawakami *et al.* (2000) to grow (Ga,Mn)As-DFH. It is important to use this method to control the positions of Mn atoms in the growth direction. The growth chamber was a UHV Varian/EPI Gen-II system. By starting with a GaAs(100) substrate, a 250 nm GaAs buffer layer was grown on top at 580 °C. There was a 2×4 reconstruction detected by RHEED at this stage. The substrate is then cooled to 300 °C in an As_2 flux. Then the RHEED pattern changed to a $c(4 \times 4)$ reconstruction. The actual growth temperature was between 240 and 280 °C. Another 100 nm GaAs buffer layer was grown to allow the substrate temperature to equilibrate with the Ga shutter open. The RHEED pattern showed a 1×1 reconstruction. The DFH is then grown by alternately opening the Ga and Mn shutters while holding a constant As_2 flux. A valve on the cracking source maintained at 700 °C controlled the flux. The ratio of As/Ga was about 25. During the deposition of Mn atoms, the RHEED pattern changed to a 1×2 reconstruction, then the 1×1 pattern was recovered when Ga was deposited. The RHEED oscillations were monitored to determine the deposition rates, about 0.5 ML/s for GaAs and 0.06 ML/s for MnAs. The DFH is capped by a GaAs layer with about 40 nm thickness. During the growth, measurements of the optical absorption spectrum of the GaAs substrate were used to monitor its temperature. The temperature of the substrate was also measured by a thermocouple. Its reading was 30 to 40 °C lower than the actual temperature due to the radiative heating of the Ga cell. The final samples have about 100 repetitions of 10 ML of GaAs and 0.5 ML of Mn. It is labeled as $(10/0.5)_{100}$.

The $(10/0.5)_{100}$ sample was subjected to X-ray diffraction analysis. The plot of X-ray counts as a function of 2θ is shown in Fig. 5.29 for the sample grown at 280 °C. The DFH(0) peak is at $2\theta = 66°$ indicating a 0.4% expansion of the lattice constant along the growth direction. The two small peaks labeled as ± 1 reflect the periodicity of the DFH. Transmission electron microscopy measures the image of the cross section at the interface. The high-resolution TEM images show that Mn is distributed over a thickness of 3–5 ML for all growth temperatures.

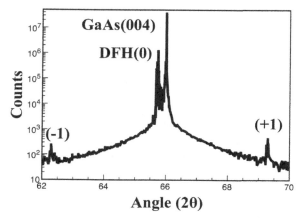

Fig. 5.29. X-ray diffraction result of $(10/0.5)_{100}$ DFH grown at $280°$C (Kawakami *et al.*, 2000).

Luo *et al.* (2002) used MBE to grow (Ga,Mn)As-DFH with a temperature variation in growing buffer layers. The growth started with a GaAs buffer layer of 200 Å on GaAs(100) substrate at $580\,°$C. The substrate temperature was then brought down to $275\,°$C and a layer of low-temperature GaAs was grown to a thickness of a few hundred Å. Finally, (Ga,Mn)As-DFH was prepared, first by MBE to grow GaAs on buffer layers and then atomic layer epitaxy (ALE) for Mn layers. They used RHEED to monitor the growth, in particular for the MnAs precipitates. Typically, samples of 9 layers of GaAs with 0.2–0.5 ML Mn were obtained.

5.5.1.2. *Growth and characterization of (Ga,Mn)Sb-DFH*

Chen *et al.* (2002) used MBE to grow (Ga,Mn)Sb-DFH on a GaAs(100) substrate. To avoid the large lattice-constant mismatch (7.5%) between the GaSb and GaAs, a 500 nm GaSb buffer layer was grown on a GaAs substrate. RHEED was used to monitor the growth. The samples have about 50 periods of 0.5 ML of Mn atoms with various GaSb thickness. TEM image of a $(12/0.5)_{50}$ DFH is shown in Fig. 5.30. The light and dark lines are clearly identifiable. The dark lines are those containing the Mn atoms.

5.5.1.3. *Physical properties of (Ga,Mn)As-DFH*

The properties probed by experimental means are the magnetic properties and T_C on (Ga,Mn)As-DFH. For (Ga,Mn)Sb-DFH, the magnetic and transport properties have been measured.

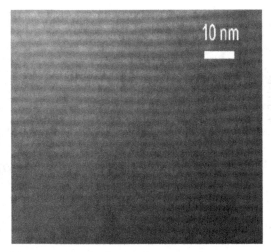

Fig. 5.30. The TEM image of a $(12/0.5)_{50}$-DFH; the Mn atoms are contained in dark lines (Chen *et al.*, 2002).

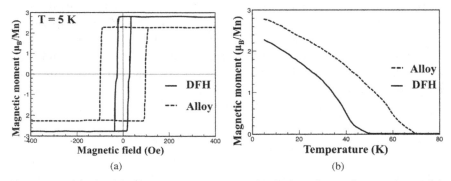

(a)

(b)

Fig. 5.31. (a) The SQUID hysteresis loop for a $(10/0.5)_{100}$ (Ga,Mn)As-DFH, and (b) the remanent magnetization as a function of temperature (Kawakami *et al.*, 2000).

Magnetic properties of (Ga,Mn)As-DFH The magnetic properties were measured by Kawakami *et al.* (2000) using a SQUID magnetometry. In Fig. 5.31, the hysteresis loop of a $(10/0.5)_{100}$ DFH grown at 280 °C is compared to the one for a dilute alloy $Ga_{0.949}Mn_{0.051}As$. The experiment was carried out at 5 K. The magnetic field is along the [100] in-plane axis — the easy axis. The results are presented with the contribution of a linear background from GaAs being subtracted. Solid lines are for the DFH, while the dashed lines are for the alloy. There is a 100% remanence showing the FM

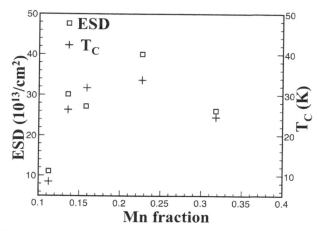

Fig. 5.32.   Effective spin density (ESD) and Curie temperature T_C of a (Ga,Mn)As-DFH as a function of Mn concentration in the layer (Luo *et al.*, 2002).

ordering along an easy magnetization axis. The temperature dependence of the remanent magnetization is shown in Fig. 5.31(b). The magnetization M was initially achieved by applying an external magnetic field of 1000 Oe at 5 K. Then, measurements of $M(T)$ were carried out at zero field as temperature increases. At about and beyond 50 K, the magnetic moment of the DFH is zero. The alloy magnetization persists up to about 70 K.

SQUID magnetometry was used by Luo *et al.* (2002) to measure the magnetizations of the samples. There are some differences between the samples grown by Luo *et al.* (2002) and those synthesized by Kawakami *et al.* (2000). Instead of using M, Luo *et al.* defined the term effective spin density (ESD) from the saturation magnetization expressed in terms of the total spin aligned by the field per unit area. The experiments were carried out at 5 K with magnetic fields up to 55 kGauss (kG). ESD plotted as a function of Mn fraction is shown in Fig. 5.32, where ESD exhibits a maximum value of 40 in units of $10^{13}/\text{cm}^2$ at Mn fraction 0.25. At Mn fraction 0.45, ESD is only about $10 \times 10^{13}/\text{cm}^2$.

Magnetic properties of (Ga,Mn)Sb-DFH   For (Ga,Mn)Sb-DFH, Chen *et al.* (2002) also used a SQUID magnetometer to measure its magnetization. All the samples show hysteresis loops up to 400 K — the limiting temperature of their SQUID magnetometer. The results of the magnetic moment at different temperatures are given in Fig. 5.33. These authors concluded that the FM properties exhibited at RT are contributed by

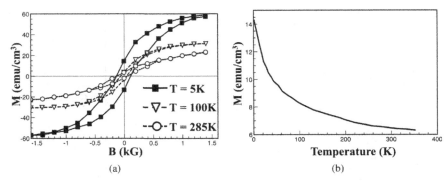

Fig. 5.33. (a) Hysteresis loops of (Ga,Mn)Sb measured at different temperatures. (b) Magnetization as a function of temperature (Chen *et al.*, 2002).

DFH, not by any MnSb precipitates. According to Abe *et al.* (2000), the RT ferromagnetism has been observed in GaMnSb if there are MnSb precipitates. But such samples show temperature-independent hysteresis loops with the coercive field at RT to be the same value as at 5 K. The results shown in Fig. 5.33 do not display such features in GaMnSb with precipitates.

The temperature dependences of magnetization given in Figs. 5.31(b) and 5.33(b) are quite different. It is difficult to compare shapes of the magnetic moment between (Ga,Mn)As- and (Ga,Mn)Sb-DFHs. On the other hand, the magnetizations $M(T)$ shown in both figures expose the difference between these two types of DFHs. $M(T)$ in (Ga,Mn)As-DFH is not zero at $T > 50$ K, while $M(T)$ is zero for (Ga,Mn)Sb-DFH.

T_C of GaAs- and GaSb-based DFH

For (Ga,Mn)As-DFH, Kawakami *et al.* (2000) reported measurements of T_C vs. GaAs thickness (Fig. 5.34). All the samples were grown at 280 °C. There is a factor of two difference between values of T_C for samples having 0.5 ML of MnAs coverage (50 K) and the one with 0.25 ML of MnAs coverage (22 K) on 10 to 15 layers of GaAs. T_C of the 0.5 ML case decreases from 50 K as the GaAs layer thickness increases. It reaches a minimum (35 K) at about 50 layers of GaAs. Then, the value attained a maximum (40 K) at 100 layers. At 200 layers, T_C reduces to 33 K. These authors remarked that T_C is sensitive to the growth temperature.

The (Ga,Mn)As-DFH samples obtained by Luo *et al.* (2002) reach the maximum T_C of 37 K at 0.25 ML of Mn (Fig. 5.32). This value is higher than the one measured by Kawakami *et al.*. However, with samples at 0.40 ML

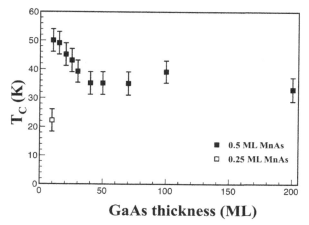

GaAs thickness (ML)

Fig. 5.34. T_C as a function of GaAs layer thickness for $(10/0.5)_{100}$ and $(10/0.25)_{100}$ DFHs (Kawakami *et al.*, 2000).

of Mn, the value determined by Kawakami *et al.* is higher by about 10 K. Whether the discrepancy can be attributed to the thicker GaAs buffer layer used by Luo *et al.* remains to be answered.

T_C of DFHs based on GaSb were estimated from the measurements of hysteresis loops by Chen *et al.* (2002). Their estimated value is higher than 400 K. The result seems to be consistent with the measurements on MnAs and MnSb in the NiAs structure. The former has a T_C at 310 K and the latter is at 580 K. We summarize the values of T_C in these known III-V compound-based DFHs in Table 5.6.

Transport properties of (Ga,Mn)As- and (Ga,Mn)Sb-DFH   Two types of measurements were carried out by Luo *et al.* on (Ga,Mn)As-DFH and by Chen *et al.* on (Ga,Mn)Sb-DFH: MR and anomalous Hall effect (AHE). The measurements were carried out using the van der Pauw configurations in a cryostat having a 170 kG (kGauss) superconducting magnet and Hall bar configuration. None of the (Ga,Mn)As-DFH samples exhibit any metallic features. The sheet resistance of a sample, (9/0.5), is fitted well with

$$\ln R = \ln R_0 - (T/T_0)^{\frac{1}{2}}, \qquad (5.5)$$

where T_0 is 61 K and R_0 is 850 Ω. This exponential dependence of R with respect to T is known for materials such as an n-type δ-doped GaAs where the conduction is contributed by a variable range of hopping. On the other

Table 5.6. T_Cs of (Ga,Mn)As- and (Ga,Mn)Sb-DFHs. The thickness of the substrate or buffer layers is represented in monolayers (ML).

DFH	Thickness (ML)	Growth temp. (°C)	T_C (K)	Reference
		(Ga,Mn)As		
$(10/0.5)_{100}$	10–15	280	50	Kawakami *et al.* (2000)
$(10/0.5)_{100}$	100	280	40	Kawakami *et al.* (2000)
$(10/0.5)_{100}$	200	280	33	Kawakami *et al.* (2000)
$(10/0.25)_{100}$	100	280	22	Kawakami *et al.* (2000)
$(10/0.5)_{100}$		260	19	Kawakami *et al.* (2000)
$(10/0.5)_{100}$		240	5	Kawakami *et al.* (2000)
$(9/0.25)$	>100	275	33	Luo *et al.* (2002)
$(9/0.4)$	>100	275	17	Luo *et al.* (2002)
		(Ga,Mn)Sb		
$(various/0.5)_{50}$		>400		Chen *et al.* (2002)

hand, all the (Ga,Mn)Sb-DFHs are metallic based on the MR results. This is in distinct contrast to the (Ga,Mn)As-DFH case.

AHE is able to probe the information about the interaction between mobile carriers and local spin moment of magnetic elements at low external magnetic field. The expression of the transverse (perpendicular to the external applied electric field) resistivity ρ_{xy} contains a term induced by the magnetization M.

$$\rho_{xy} = R_0 B + 4\pi R_a M, \qquad (5.6)$$

where R_o is the normal Hall coefficient, B is the strength of the external magnetic field, and R_a is the anomalous Hall coefficient. The first term is the normal Hall effect. The second term defines the AHE.

In Fig. 5.35, the sheet and Hall resistances of (Ga,Mn)As-DFH are shown. For (Ga,Mn)As-DFH, the sheet resistance decreases with temperature. The shapes do not change significantly as the temperature varies from 12 to 52 K. They are peaked at $B = 0$. With the four-terminal van der Pauw technique, the Hall resistance R_{xy} is equivalent to the Hall resistivity ρ_{xy} (Metalidis and Bruno, 2006). The magnitude of R_{xy} (R_{Hall}) decreases with temperature. R_{xy} changes from negative to positive value in a very narrow ($\sim$20 kG) region when the positive external magnetic field reverses its sign. The question of the physical origins causing the temperature behavior, the double structures in the sheet resistances, and their small R_{xy}

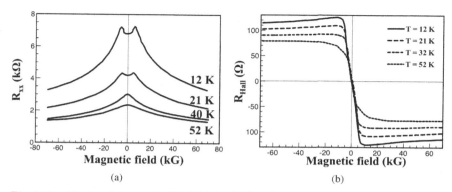

Fig. 5.35. Sheet resistance of a (Ga,Mn)As-DFH at different temperatures as a function of external magnetic field (a), and Hall resistance of a (Ga,Mn)As-DFH at different temperatures as a function of external magnetic field (b) (Luo *et al.*, 2002).

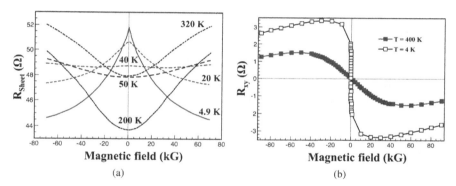

Fig. 5.36. (a) Sheet resistance of a (Ga,Mn)Sb-DFH at different temperatures as a function of external magnetic field, and (b) the Hall resistivity at $T = 4$ and $400\,\mathrm{K}$ as a function of external magnetic field (Chen *et al.*, 2002).

values compared to the large ones (one order of magnitude larger) is not answered.

Since the above-mentioned question is not answered, we just point out the difference of the magnetotransport properties of the two (Ga,Mn)-pnictide-DFHs. For (Ga,Mn)Sb-DFH, both the sheet and Hall resistances behave quite differently from the case of (Ga,Mn)As-DFH. Let us first compare the sheet resistances shown in Figs. 5.35(a) and 5.36(a). At low temperature $(T < 40\,\mathrm{K})$, the shapes are somewhat similar. However, the magnitudes differ by more than a factor of 10 at about $20\,\mathrm{K}$. In addition, those of (Ga,Mn)Sb-DFHs do not exhibit the double structure. The shapes of R_{sheet} shown in Fig. 5.36 are also changed drastically at higher temperature ($T \geq 40\,\text{K}$). Instead of decreasing, the sheet resistances increase as the magnetic field increases positively and decreases negatively. Two qualitative features of R_{xy} in both DFHs are in agreement: (i) the decrease in magnitudes of the resistance as the temperature increases; (ii) the sharp reverse of R_{xy} (change from positive to negative) vs. external magnetic field at low temperature. At high temperature ($T = 400\,\text{K}$), R_{xy} of (Ga,Mn)Sb-DFH shows a smooth transition from positive to negative as the magnetic field changes from negative to positive. No such behavior was detected for the temperature range between 21 and 32 K, just below the T_C of (Ga,Mn)As-DFH.

5.5.2. *Theory*

5.5.2.1. *GaAs-based DFH*

Structure Immediately after publication of the experimental work by Kawakami *et al.* (2000), Sanvito and Hill (2001) reported their theoretical study of an $(N_{\text{GaAs}}/1)_\infty$ (Ga,Mn)As-DFH, where N_{GaAs} is the number of GaAs layers and ∞ stands for a superlattice. They used SIESTA algorithm with the LSDA (Sánchez-Portal *et al.*, 1997). The experimentally determined lattice constant of GaAs, 5.65 Å, was used. These authors also examined antisite defects.

Electronic and magnetic properties The band structures near E_F for the majority- and minority-spin channels of a $(15/1)_\infty$ (Ga,Mn)As-DFH are shown in Figs. 5.37(a) and (b), respectively. The directions of **k** are in the plane perpendicular to the GaAs layers. X_1 is at the edge of the square BZ and X_2 is located at the corner, the diagonal, of the square. The ↑ spin states show metallic behavior while the ↓ spin channel exhibits a gap of 0.65 eV. The DFH is a two-dimensional HM. The band dispersions along the two directions show anisotropy. There is no identification of the origins of those states. The widths of those bands intercepted by E_F become narrower as the separation between Mn layers increases.

By comparing the FM and AFM configurations of the Mn atoms, the authors found that the FM configuration is more favorable. The coupling strength is stronger in a DFH than in a random $Ga_{1-x}Mn_xAs$ alloy. The energy difference Δ_{FA} between ferro- and anti-ferromagnetic configurations

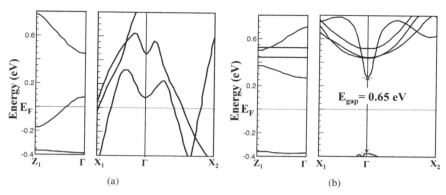

Fig. 5.37. Band structures near E_F of the (a) majority- and (b) minority-spin channels for a $(15/1)_\infty$ (Ga,Mn)As-DFH (Sanvito and Hill, 2001).

is 515 meV in the $(15/1)_\infty$ structure while Δ_{FA} in the random alloy is only 160 meV. For DFHs, this energy difference depends weakly on the GaAs thickness for small N. When $N = 4$, 6, and 8, the values of Δ_{FA} are 531, 533, and 515 meV, respectively.

Defects Sanvito and Hill investigated the effects of As antisites, that is to have As atoms occupy Ga sites. In the supercell model of $(11/1)_\infty$ they replaced half of the Ga atoms by the As atoms in the layer close to the MnAs layer. The FM phase is still favored with Δ_{FA} to be 70 meV. They cited the Zener model in which the coupling between the hole spin and local spin moment at the Mn atom causes the long-range FM behavior.

Transport properties The ballistic transport properties of (Ga,Mn)As-DFHs were investigated by Sanvito and Hill. They calculated the current in the Mn plane termed as CIP and the one perpendicular to the Mn plane, CPP. The Landauer–Büttiker formula (Büttiker *et al.*, 1985) was used to calculate the conductance of each of the two spin channels.

$$\Gamma_\sigma = \frac{2e}{h} \sum_k Tr\left[t_\sigma(k)t_\sigma^\dagger(k)\right], \tag{5.7}$$

where Tr is the notation of trace, t_σ is the transmission matrix for spin σ. The transmission matrix is the scattering matrix with matrix elements between the incoming and outgoing waves.

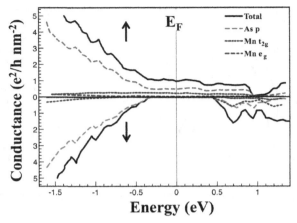

Fig. 5.38. Total and partial spin-polarized CIP conductances of a $(15/1)_\infty$ DFH (Sanvito and Hill, 2001).

The results of the calculated spin-polarized CIP conductances of a $(15/1)_\infty$ DFH are shown in Fig. 5.38. The $\uparrow$ spin channel of the CIP conductance is finite in the region around E_F. From the partial CIP conductance, the contributions are dominated by p-states of the As atoms. The t_{2g} states of the Mn atoms contribute less. The e_g states of the Mn atom hardly make a significant contribution. These results are consistent with the crystal field and hybridization effects. The contributions from As p-states and Mn t_{2g} states are the consequence of hybridization. With the tetrahedral symmetry, the e_g states are the nonbonding states. They do not hybridize strongly with p-states of the neighboring As atoms (Fig. 1.10). The $\downarrow$ spin states do not contribute to any conductance near E_F. The DFH is a 2-D HM. Because of the thick GaAs (15) layers, it is not expected to have any CPP conductance.

5.5.2.2. *Si-based DFH*

Structure From what we have discussed so far, it is evident that enormous efforts have been devoted to the study of HMs. However, at this moment, we still do not have any realization of spintronic devices using HMs. The major obstacle is the growth of Heusler alloys, oxides, and TM pnictides. Then, there is the T^* ($\sim 88\,\mathrm{K}$) at which the half-metallicity disappears in a Heusler alloy due to spin-flip transitions and the fact that P vanishes precipitously as the temperature approaches RT.

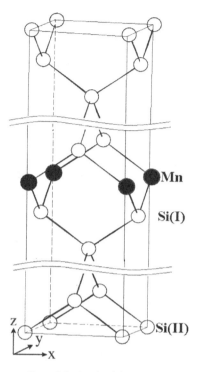

Fig. 5.39. A supercell model of a $(31/1)_\infty$ DFH (Qian *et al.*, 2006a).

Qian *et al.* (2006a) proposed a DFH by doping a δ-layer of Mn atoms in Si (Si-based DFH), motivated by the fact that Si technologies are the most mature among the semiconductors so the growth problems can possibly be eliminated. They used a supercell model of a $(31/1)_\infty$ DFH shown in Fig. 5.39. The plane wave pseudopotential method, and the GGA exchange-correlation within DFT, were used to calculate the density of states of Si-based DFH for the investigation of whether there is a T^* in this DFH.

Effect of lattice relaxation The presence of 1.0 ML of Mn in 31 layers of Si necessitates examining various properties affected by the lattice relaxation. The effects on the magnetic moment, DOS at E_F in the conducting channel, Δ_{FA} of the Mn atoms, and energy gaps are summarized in Table 5.7. The magnetic moment is 3.0 μ_B with finite DOS at E_F ($N(E_F)$) for the majority-spin channel. E_{FM} and E_{AFM} are the total energies for the FM and AFM configurations, respectively. E_g is the energy gap in the

Table 5.7.   Comparison of the unrelaxed and relaxed $(31/1)_\infty$-DFH. m is the magnetic moment per Mn atom, $N(E_F)$ is the DOS of the conducting channel at E_F, $\Delta_{\mathrm{FA}} = E_{\mathrm{FM}} - E_{\mathrm{AFM}}$, and E_g is the energy gap.

Structure	m (μ_B)	$N(E_F)$ (states/eV-cell)	Δ_{FA} (meV)	E_g (eV)
Unrelaxed	3.0	1.06	−523.91	0.20
Relaxed	3.0	1.25	−442.38	0.25

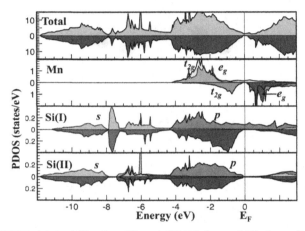

Fig. 5.40.   TDOS (states/eV-unit cell) and PDOS (states/eV) for a $(31/1)_\infty$ DFH. PDOS are for the nearest- and second-nearest-neighbor Si atoms (Qian *et al.*, 2006a).

minority-spin states. Relaxation of the atoms inside the unit cell maintains the magnetic moment/Mn atom and improves $N(E_F)$ and E_g. E_{FM} is reduced but the FM phase is still favorable.

Density of states   The spin-polarized DOS indicates whether a sample is an HM. The calculated results are shown in Fig. 5.40. The top panel shows the total DOS. The second one is for a Mn atom. The third and fourth panels are for the Si atoms located at the nearest- and second-nearest-neighbor Si atoms of the Mn atom. At E_F, TDOS shows that the majority-spin channel is metallic while the minority states exhibits insulating behavior. The gap is 0.25 eV. Combining with the integer spin moment per unit cell ($3\,\mu_B$/unit cell), the DFH is an HM.

   The PDOS enable one to identify states around E_F. With the energy window between $E_F - 1.0$ and $E_F + 2.0$ eV, the majority-spin states are primarily the hybridized p-states of the nearest neighbor Si atom (Si(I)) to the Mn atom and the t_{2g} states of the Mn atom (the second and the third panels). As will be shown later, these are part of the antibonding states.

The strongest region of the hybridization is at 2.6 eV below E_F. The e_g states of the Si(I) atom are centered around -2.5 eV with a width of about 1.0 eV. In the range between $E_F - 1.0$ and $E_F + 2.0$ eV, the second neighbor Si(II) atom has hardly any significant contributions. Between $E_F - 2.0$ and $E_F - 4.0$ eV, those p-states have a uniform distribution.

In the minority-spin channel, those states just below the valence band maximum are the p(Si(I))-d(Mn(t_{2g})) hybridized bonding states. E_F is located near the top of the valence bands. The e_g states of the Mn atom are not occupied. The conduction band minimum is contributed by the low energy end of the e_g states. Neither Si(I) nor Si(II) have contributions to the states at the conduction band minimum.

Band structures and Fermi surfaces The band structures near E_F of the ↑ and ↓ spin channels are plotted in Figs. 5.41(a) and (b), respectively. Γ–Z is in the direction perpendicular to the layers. The top of the valence band along this direction is relatively flat as compared to lower energy bands. There are gaps at the Γ- and Z-points. Along this direction, there is no conduction. The R-point is at the corner of the two-dimensional BZ. The X-point is along the [100] direction. The occupied bands closest to E_F at these two points are derived from the antibonding states at the Γ-point. Therefore, the Γ-point is expected to be surrounded by a hole Fermi surface while the R-point should have electron pockets. The Fermi surfaces thus can have both electron and hole surfaces in the two-dimensional BZ and are shown in Fig. 5.42. In Fig. 5.42(a) the contour surrounding the

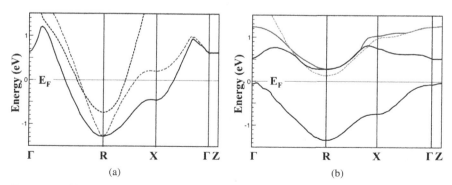

Fig. 5.41.   (a) The band structure of the majority-spin states and (b) the band structure of the minority-spin channel (Qian *et al.*, 2006a).

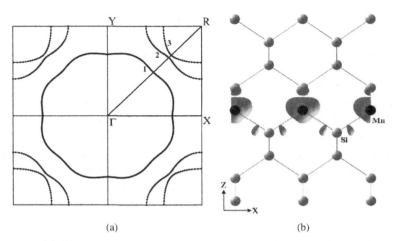

(a)   (b)

Fig. 5.42. (a) The Fermi surface in the X–Y section of the two-dimensional BZ, and (b) the hole charge distribution for the states near E_F (Qian *et al.*, 2006a).

Γ-point is the hole surface. The curves at the R-point are the boundaries of electron surfaces. The hole charge distribution for states near E_F is shown in Fig. 5.42(b).

T* in Si-based DFH   According to recent experiments on dilutely doped Mn in Si, $Mn_x Si_{1-x}$ alloys, T_C can be above 400 K (Zhang *et al.*, 2004; Bolduc *et al.*, 2005; Bandaru *et al.*, 2006). It is especially encouraging because Bandaru *et al.* (2006) verified that the high T_C and ferromagnetism are intrinsic properties, not due to clustering of Mn atoms. A relevant issue, the question of whether this DFH can have T_C at or above RT, needs to be addressed in order to determine whether this DFH will be a potential spintronic material. In Fig. 5.43, we make use of the one shown in Section 3.6.1.2 (Fig. 3.12). E_F of the Heusler alloy is denoted by "E_F of HA" and is located right below the conduction band edge (CBMin) in the $\downarrow$ spin channel.

We now focus on the Si-based DFH. Its E_F is explicitly labeled and is located near the VBM of the $\downarrow$ spin states. The gap of the insulating minority-spin states is denoted by $\Delta_\downarrow$. The important gap is labeled as δ which is measured from E_F to CBMin in the minority-spin channel. Its value is approximately 0.25 eV (Table 5.7) which is about 10 times larger than RT. Therefore, the probability is drastically reduced for an electron in the majority-spin channel to make a spin-flip transition from E_F to CBMin of the minority-spin channel. However, one can ask the following question: what about the possibility for an electron from the VBM to

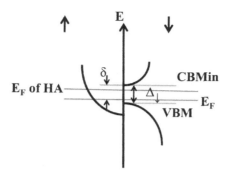

Fig. 5.43. Schematic diagram of the DOS of an HM with the majority-spin channel showing metallic behavior. The Fermi energy of the Heusler alloy is now indicated as "E_F of HA" and is right below CBMin. The Fermi energy of the Si-based DFH is labeled as E_F and is located close to the valence band maximum (VBM) of the DFH. δ is the energy difference between CBMin and E_F. $\Delta_\downarrow$ is the fundamental gap of the insulating channel.

make a spin-flip transition to a state at E_F? As we see from the PDOS (Fig. 5.40), the valence band is dominated by p-states of Si(I). Consequently, one expects the corresponding charge density to be concentrated near the Si atom. The overlap between the Si(I) p-states and Mn d-states will be small. As shown in Fig. 5.40, the hole states near E_F have their charge distributions concentrated on the Mn atoms. The transition matrix element reduces the probability of this kind of spin-flip transition. The conclusion is that there will not be a T^* in the Si-based DFH. Therefore, it can be a promising material for making spintronic devices.

5.6. One-dimensional Half-metals

Even though DFHs behave as two-dimensional systems, their basic structures are still three-dimensional. To explore the possibility of finding half-metallicity in low-dimensional systems, Dag *et al.* (2005) investigated one-dimensional systems by doping TM elements in carbon wires, $C_n(TM)$. They used a pseudopotential method within DFT with spin-polarized GGA exchange-correlation.

Model The wires are modeled by a tetragonal supercell with $a = b = 10$ Å and c depends on the length of the carbon wires. Sometimes, c is set to be twice the length of the wires if Peierls instability (Peierls, 1955) is the concern. While they covered different n's and many TM elements, such as

Table 5.8. Summary of calculated total energy difference ΔE_T between the spin-polarized and spin-unpolarized cases, optimized value of c, magnetic moment/unit cell (M), and the sample type: semiconductor (S) or HM. For HMs, the energy gap is also given.

Compound	ΔE_T (eV)	c (Å)	M (μ_B)	Type	Gap (eV) ↑	↓
CCr	−1.8	3.7	2.0	S	0.7	1.0
C_2Cr	−2.8	5.2	4.0	HM		3.3
C_3Cr	−3.0	6.5	4.0	HM	0.4	
C_4Cr	−3.0	7.9	4.0	HM		2.9
C_5Cr	−2.5	9.0	4.0	HM	0.6	
C_6Cr	−3.1	10.3	4.0	HM		2.4
C_7Cr	−2.5	11.6	4.0	HM	0.5	

Cr, Fe, Mn and Ti, they reported mainly on C_nCr with n ranging from 2 to 7. These results are summarized in Table 5.8.

Half-metallic properties and magnetic moments Except the CCr compound, the other compounds with $n \geq 2$ are HMs. These systems favor spin polarization and have lower energies than the corresponding unpolarized cases. The values of the magnetic moment/unit cell ($4.0\,\mu_B$) are larger than the one for MnC with larger lattice constant. A simple explanation is that each of the two neighboring C atoms transfers one electron from the Cr atom to doubly occupy its p-orbital. It is interesting to note that the half-metallic behavior depends on n. The metallic feature alternates between the majority- and minority-spin states as n increases from 2 to 7. For even n, the majority-spin channel is metallic. In addition to the electronic properties, these authors also investigated the stability of the wires and the effects of spin–orbit interaction. Based on their findings, they suggested a way to grow these HMs.

Spin-polarized band structure and DOS The spin-polarized band structures and density of states of C_3Cr and C_4Cr are shown in Fig. 5.44. The band structures are plotted along the direction of the wire, z-direction (Γ–Z). The solid lines are for the majority-($\uparrow$) spin states and dashed lines are for the minority-($\downarrow$) spin channel. E_F is set to be zero. $v\sigma$ and $m\sigma$ are, respectively, the highest valence and lowest conduction bands for the spin-σ channel. The two $m\sigma$ bands are intercepted by E_F having different spin orientations. $c\sigma$ is the next conduction band with energy larger than the $m\sigma$ state. If the TM element has the symmetric environment, $n = 4$,

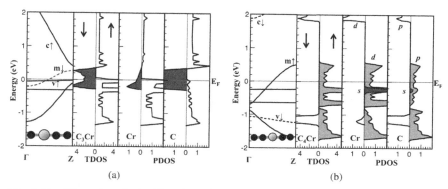

Fig. 5.44. The spin-polarized band structures and density of states of (a) C_3Cr and (b) C_4Cr (Dag *et al.*, 2005).

a band in the majority-spin channel is intercepted by E_F. On the other hand in the asymmetric cases, $n = 3$, the band intercepted by E_F belongs to the minority-spin states. Looking closer from the bond lengths, double bonds are formed between all atoms for $n = 3$, where the bond lengths in Å are distributed as C-1.28-C-1.28-C-1.95-Cr. For $n = 4$, triple and single bonds form alternately between C atoms, and a longer bond between the Cr atom and its neighbors. The distributions are C-1.25-C-1.33-C-1.25-C-2.1-Cr. The different bond lengths between the Cr and C atoms result in different band structures.

From the PDOS, the $m\sigma$ bands are formed by Cr-3d states hybridized with 2p-state of the neighboring C atoms. The $v\sigma$ band is composed of s-states and d-states of the Cr atom. The unoccupied $c\sigma$ band is contributed by the hybridized antibonding C p-states and Cr d-states.

Effect of spin–orbit interaction To accurately investigate the spin–orbit interaction, the total energy difference was calculated using WIEN2K. It is found that the effect is small, -7.9 meV with the spin–orbit interaction relative to the non-spin–orbit case. The TDOS with and without spin–orbit interaction for the case C_3Cr are shown in Fig. 5.45. E_F is set at zero. The shift of the states in the majority-spin states due to the spin–orbit interaction does not destroy the half-metallicity.

Stability These authors carried out an extensive search of various structures. Local minima of the total energy were probed by optimizing the structure starting from a transversely displaced chain of atoms at various lattice constants. The linear chain structure has been found to be stable

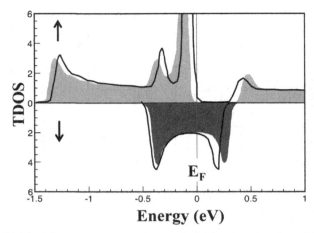

Fig. 5.45.   TDOS of the two spin channels with (solid lines) and without (shaded areas) spin–orbit interaction for C_3Cr (Dag *et al.*, 2005).

and is more favorable than a zig-zag structure. They also calculated the phonon frequencies Ω. They are all positive. For $n = 3$, $\Omega_{TO}(\mathbf{q} = 0) = 89$, 92, 411 cm^{-1} and $\Omega_{LO}(\mathbf{q} = 0) = 421$, 1272, 1680 cm^{-1}. The corresponding frequencies for $n = 4$ are 13, 71, 353, 492 cm^{-1} and 489, 1074, 1944, 2102 cm^{-1}. However, for $n = 9$ some of the frequencies become negative. To assure the stability of the chains with small n (<9), they also carried out *ab initio* molecular dynamics simulations with T set between 750 and 1000 K using the Nosé thermostat. The atoms were displaced in random directions. The results confirm the stability of the chains at small n.

There are still the possibilities of the Peierls instability and strain issues. These authors expanded the models by a factor 2 and calculated the band structures. The splitting of bands at E_F should manifest the effect of the Peierls instability. There is no splitting of metallic bands at E_F in C_nCr chain structures. Axial strains were applied to examine whether strain can destroy half-metallicity. The half-metallicity is robust for $n = 4$ under $\epsilon_z = \pm 0.05$. However, for $n = 5$, its half-metallic properties remain only at $\epsilon_z \leq 0.05$. With $\epsilon_z = -0.05$, C_5Cr is transformed to an FM metal. At $\epsilon_z = 0.10$, C_3Cr is a semiconductor, while it becomes an FM metal with the magnetic moment of $3.1\,\mu_B$/unit cell when $\epsilon_z = -0.10$.

Growth   In Fig. 5.46(a), the doping path gives the energetics of a process of doping Cr in C_5. The steps are indicated as A, B, C, and D. From A to D, there is a barrier of 1.86 eV. This is considered to be a large barrier to

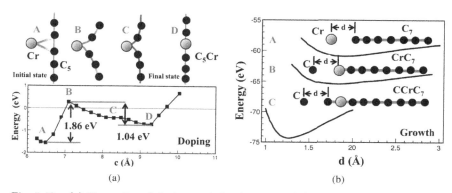

Fig. 5.46. (a) Energetics of doping path for C_5Cr, and (b) a possible path of growing C_7Cr wire (Dag *et al.*, 2005).

overcome. Therefore, the doping process is not a viable method to grow a half-metallic wire. The growth path in Fig. 5.46(b) suggests a possible way to grow a C_7Cr wire without a large barrier. The process A first shows the growth of a C_7 chain, then a Cr atom is deposited at one end of the C_7. The processes of growing C_7Cr wire by adding more C atoms to one end of the Cr atom are indicated as B and C. From the interaction energy with distance d there is no significant barrier.

Appendix A

Anisotropic Magnetoresistance

InAnisotropic magnetoresistance (AMR) a ferromagnetic metal, the resistivity (the resistance multiplied by cross section and divided by height for a cylindrical sample) can change when the magnetization (M) of the metal changes under (i) change of temperature and/or (ii) change of direction of applied field. Anisotropic magnetoresistance (AMR) arises under condition (ii). Experimentally, one measures the resistivities, $\rho_{\parallel}$ and $\rho_{\perp}$ vs. applied magnetic field $\mathbf{H}_{ext}$, where $\rho_{\parallel}(\rho_{\perp})$ is the measured resistivity with the applied field parallel (tranverse) to the direction of current. In an AMR material, $\rho_{\parallel}$ typically increases with applied field while $\rho_{\perp}$ typically decreases (see Fig. A.1). AMR is defined as $\Delta\rho = \rho_{\parallel}(A) - \rho_{\perp}(B)$ where $\rho_{\parallel}$ (A) and $\rho_{\perp}$ (B) are the measured resistivities at points A and B, where each reach saturated values vs. $\mathbf{H}_{ext}$ (Fig. A.1). The normalized AMR is defined as $\Delta\rho/\rho_{av}$, where $\rho_{av} = (1/3)\rho_{\parallel} + (2/3)\rho_{\perp}$. The advantage of using the normalized AMR is that it is dimensionless. Thus, there is no need to know the dimension of the sample.

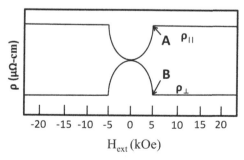

Fig. A.1. Schematic representation of magnetoresistivity of a ferromagnetic metal vs. $\mathbf{H}_{ext}$.

Bibliography

Abe, E., Matsukura, F., Yasuda, H., Ohno, Y. and Ohno, H. (2000). Molecular beam epitaxy of III-V diluted magnetic semiconductor (Ga,Mn)Sb, *Physica E: Low-dimensional Systems and Nanostructures* **7**, p. 981.

Akai, H., Akai, M. and Kanamori, J. (1985). Electronic structure of impurities in ferromagnetic iron. II. 3d and 4d impurities, *J. Phys. Soc. Jpn.* **54**, p. 4257.

Akai, H. and Dederichs, P. H. (1993). Local moment disorder in ferromagnetic alloys, *Phys. Rev. B* **47**, p. 8739.

Akinaga, H., Manago, T. and Shirai, M. (2000a). Material design of half-metallic zinc-blende CrAs and the synthesis by molecular-beam epitaxy, *Jpn. J. Appl. Phys. Lett.* **39**, p. L1118.

Akinaga, H. and Mizuguchi, M. (2004). Zinc-blende CrAs/GaAs multilayers grown by molecular-beam epitaxy, *J. Phys.: Condens. Mat.* **16**, p. S5549.

Akinaga, H., Mizuguchi, M., Ono, K. and Oshima, M. (2000b). Room-temperature thousandfold magnetoresistance change in MnSb granular films: Magnetoresistive switch effect, *Appl. Phys. Lett.* **76**, p. 357.

Albert, F. J., Katine, J. A., Buhrman, R. A. and Ralph, D. C. (2000). Spin-polarized current switching of a Co thin film nanomagnet, *Appl. Phys. Lett.* **77**, p. 3809.

Alonso, J. A., Martinez-Lope, M. J., Casais, M. T., Macmanus-Driscoll, J. L., de Silva, P. S. I. P. N., Cohen, L. F. and Fernandez-Diaz, M. T. (1997). Non-stoichiometry, structural defects and properties of $LaMnO_{3+\delta}$ with high δ values ($0.11 \leq \delta \leq 0.29$), *J. Mater. Chem.* **7**, p. 2139.

Ambrose, T., Krebs, J. and Prinz, G. (2000). Epitaxial growth and magnetic properties of single-crystal Co_2MnGe Heusler alloy films on GaAs(001), *Appl. Phys. Lett.* **76**, p. 3280.

Andersen, O. K. (1975). Linear methods in band theory, *Phys. Rev. B* **12**, p. 3060.

Anderson, P. W. (1950). Antiferromagnetism. Theory of superexchange interaction, *Phys. Rev.* **79**, p. 350.

Anderson, P. W. (1956). Ordering and antiferromagnetism in ferrites, *Phys. Rev.* **102**, p. 1008.

Anderson, P. W. and Hasegawa, H. (1955). Considerations on double exchange, *Phys. Rev.* **100**, p. 675.

Andreev, A. (1964). Thermal conductivity of the intermediate state of superconductors, *Sov. Phys. JETP* **19**, p. 1228.

Anguelouch, A., Gupta, A., Xiao, G., Abraham, D., Ji, Y., Ingvarsson, S. and Chien, C. L. (2001). Near-complete spin polarization in atomically-smooth chromium-dioxide epitaxial films prepared using a CVD liquid precursor, *Phys. Rev. B* **64**, p. 180408(R).

Anisimov, V. I., Aryasetiawan, F. and Lichtenstein, A. I. (1997). First-principles calculations of the electronic structure and spectra of strongly correlated systems: the LDA + U method, *J. Phys.: Condens. Mat.* **9**, p. 767.

Anisimov, V. I., Elfimov, I. S., Hamada, N. and Terakura, K. (1996). Charge-ordered insulating state of Fe_3O_4 from first-principles electronic structure calculations, *Phys. Rev. B* **54**, p. 4387.

Antonov, V., Dürr, H., Kucherenko, Y., Bekenov, L. and Yaresko, A. (2005). Theoretical study of the electronic and magnetic structures of the Heusler alloys $Co_2Cr_{1-x}Fe_xAl$, *Phys. Rev. B* **72**, p. 054441.

Antonov, V., Oppeneer, P., Yaresko, A., Perlov, A. and Kraft, T. (1997). Computationally based explanation of the peculiar magneto-optical properties of PtMnSb and related ternary compounds, *Phys. Rev. B* **56**, p. 13012.

Antonov, V. N., Harmon, B. N., Antropov, V. P., Perlov, A. Y. and Yaresko, A. N. (2001). Electronic structure and magneto-optical Kerr effect of Fe_3O_4 and Mg^{2+}- or Al^{3+}-substituted Fe_3O_4, *Phys. Rev. B* **64**, p. 134410.

Asamitsu, A., Moritomo, Y., Kumai, R., Tomioka, Y. and Tokura, Y. (1996). Magnetostructural phase transitions in $La_{1-x}Sr_xMnO_3$ with controlled carrier density, *Phys. Rev. B* **54**, p. 1716.

Austin, B., Heine, V. and Sham, L. (1962). General theory of pseudopotentials, *Phys. Rev.* **127**, p. 276.

Bader, S. D. (1991). SMOKE, *J. Magn. Magn. Mater.* **100**, p. 440.

Bahr, D., Press, W., Jebasinski, R. and Mantl, S. (1993). X-ray reflectivity and diffuse-scattering study of $CoSi_2$ layers in Si produced by ion-beam synthesis, *Phys. Rev. B* **47**, p. 4385.

Baibich, M. N., Broto, J. M., Fert, A., Dau, F. N. V., Petroff, F., Etienne, P., Creuzet, G., Friederich, A. and Chazelas, J. (1988). Giant magnetoresistance of (001)Fe/(001)Cr magnetic superlattices, *Phys. Rev. Lett.* **61**, p. 2472.

Bandaru, P. R., Park, J., Lee, J. S., Tang, Y. J., Chen, L.-H., Jin, S., Song, S. A. and O'Brien, J. R. (2006). Enhanced room temperature ferromagnetism in Co- and Mn-ion-implanted silicon, *Appl. Phys. Lett.* **89**, p. 112502.

Bauer, E. (2010). Phänomenologische theorie der kristallabscheidung an oberflächen. I, *Zeitschrift für Kristallographie* **110**, p. 372.

Becke, A. D. (1988). Density-functional exchange-energy approximation with correct asymptotic behavior, *Phys. Rev. A* **38**, p. 3098.

Beeby, J. L. (1967). The density of electrons in a perfect or imperfect lattice, *Proc. R. Soc. Lond. A* **302**, p. 113.

Bezryadin, A., Verschueren, A., Tans, S. and Dekker, C. (1998). Multiprobe transport experiments on individual single-wall carbon nanotubes, *Phys. Rev. Lett.* **80**, p. 4036.

Binnig, G., Rohrer, H., Gerber, C. and Weibel, E. (1983). 7 x 7 reconstruction on Si(111) resolved in real space, *Phys. Rev. Lett.* **50**, p. 120.

Blöchl, P. E. (1994). Projector augmented-wave method, *Phys. Rev. B* **50**, p. 17953.

Block, T., Carey, M., Gurney, B. and Jepsen, O. (2004). Band-structure calculations of the half-metallic ferromagnetism and structural stability of full- and half-Heusler phases, *Phys. Rev. B* **70**, p. 205114.

Blonder, G. E., Tinkham, M. and Klapwijk, T. M. (1982). Transition from metallic to tunneling regimes in superconducting microconstrictions: Excess current, charge imbalance, and supercurrent conversion, *Phys. Rev. B* **25**, p. 4515.

Bohm, D. (1951). *Quantum Theory* (Prentice Hall Inc., Upper Saddle River, New Jersey).

Bolduc, M., Awo-Affouda, C., Stollenwerk, A., Huang, M. B., Ramos, F. G., Agnello, G. and LaBella, V. P. (2005). Above room temperature ferromagnetism in Mn-ion implanted Si, *Phys. Rev. B* **71**, p. 033302.

Bollero, A., Ziese, M., Höhne, R., Semmelhack, H. C., Köhler, U., Setzer, A. and Esquinazi, P. (2005). Influence of thickness on microstructural and magnetic properties in Fe_3O_4 thin films produced by PLD, *J. Magn. Magn. Mater.* **285**, p. 279.

Bona, G. L., Meier, F., Taborelli, M., Bucher, E. and Schmidt, P. H. (1985). Spin polarized photoemission from NiMnSb, *Solid State Commun.* **56**, p. 391.

Borca, C. N., Komesu, T., Jeong, H.-K., Dowben, P. A., Ristoiu, D., Hordequin, C., Nozières, J. P., Pierre, J., Stadler, S. and Idzerda, Y. U. (2001). Evidence for temperature dependent moments ordering in ferromagnetic NiMnSb(100), *Phys. Rev. B* **64**, p. 052409.

Bouzerar, R., Bouzerar, G. and Ziman, T. (2006). Why RKKY exchange integrals are inappropriate to describe ferromagnetism in diluted magnetic semiconductors, *Phys. Rev. B* **73**, p. 024411.

Bowen, M., Barthélémy, A., Bibes, M., Jacquet, E., Contour, J.-P., Fert, A., Ciccacci, F., Duò, L. and Bertacco, R. (2005). Spin-polarized tunneling spectroscopy in tunnel junctions with half-metallic electrodes, *Phys. Rev. Lett.* **95**, p. 137203.

Bowen, M., Bibes, M., Barthélémy, A., Contour, J.-P., Anane, A., Lemaitre, Y. and Fert, A. (2003). Nearly total spin polarization in $La_{2/3}Sr_{1/3}MnO_3$ from tunneling experiments, *Appl. Phys. Lett.* **82**, p. 233.

Braun, J. (1996). The theory of angle-resolved ultraviolet photoemission and its applications to ordered materials, *Rep. Prog. Phys.* **59**, p. 1267.

Brener, N. E., Tyler, J. M., Callaway, J., Bagayoko, D. and Zhao, G. L. (2000). Electronic structure and Fermi surface of CrO_2, *Phys. Rev. B* **61**, p. 16582.

Brown, P. J., Neumann, K. U., Webster, P. J. and Ziebeck, K. R. A. (2000). The magnetization distributions in some Heusler alloys proposed as half-metallic ferromagnets, *J. Phys.: Condens. Mat.* **12**, p. 1827.

Bruno, P. (2003). Exchange interaction parameters and adiabatic spin-wave spectra of ferromagnets: A "renormalized magnetic force theorem", *Phys. Rev. Lett.* **90**, p. 087205.

Buschow, K. H. J. and van Engen, P. G. (1981). Magnetic and magneto-optical properties of Heusler alloys based on aluminium and gallium, *J. Magn. Magn. Mater.* **25**, p. 90.

Buschow, K. H. J., van Engen, P. G. and Jongebreur, R. (1983). Magneto-optical properties of metallic ferromagnetic materials, *J. Magn. Magn. Mater.* **38**, p. 1.

Bussmann, K., Prinz, G. A., Cheng, S.-F. and Wang, D. (1999). Switching of vertical giant magnetoresistance devices by current through the device, *Appl. Phys. Lett.* **75**, p. 2476.

Büttiker, M., Imry, Y., Landauer, R. and Pinhas, S. (1985). Generalized many-channel conductance formula with application to small rings, *Phys. Rev. B* **31**, p. 6207.

Caballero, J., Park, Y., Cabbibo, A., Childress, J., Petroff, F. and Morel, R. (1997). Deposition of high-quality NiMnSb magnetic thin films at moderate temperatures, *J. Appl. Phys.* **81**, p. 2740.

Cai, Y. Q., Ritter, M., Weiss, W. and Bradshaw, A. M. (1998). Valence-band structure of epitaxially grown $Fe_3O_4(111)$ films, *Phys. Rev. B* **58**, p. 5043.

Callen, H. (1963). Green function theory of ferromagnetism, *Phys. Rev.* **130**, p. 890.

Caminat, P., Valerio, E., Autric, M., Grigorescu, C. and Monnereau, O. (2004). Double beam pulse laser deposition of NiMnSb thin films at ambient temperature, *Thin Solid Films* **453–454**, p. 269.

Carbonari, A. W., Pendl, W., Attili, R. N. and Saxena, R. N. (1993). Magnetic hyperfine fields in the heusler alloys Co_2YZ (Y=Sc, Ti, Hf, V, Nb; Z=Al, Ga, Si, Ge, Sn), *Hyperfine Interactions* **80**, p. 971.

Carra, P., Thole, B. T., Altarelli, M. and Wang, X. (1993). X-ray circular dichroism and local magnetic fields, *Phys. Rev. Lett.* **70**, p. 694.

Celotta, R., Pierce, D., Wang, G., Bader, S. and Felcher, G. (1979). Surface magnetization of ferromagnetic ni(110): A polarized low-energy electron diffraction experiment, *Phys. Rev. Lett.* **43**, p. 728.

Ceperley, D. and Alder, B. (1980). Ground state of the electron gas by a stochastic method, *Phys. Rev. Lett.* **45**, p. 566.

Chang, L. and Ploog, K. (1985). Molecular beam epitaxy and heterostructures, *NATO Adv. Sci. Inst. Ser.* **E87**, p. 719.

Chen, C. T., Idzerda, Y. U., Lin, H. J., Smith, N. V., Meigs, G., Chaban, E., Ho, G. H., Pellegrin, E. and Sette, F. (1995). Experimental confirmation of the X-ray magnetic circular dichroism sum rules for iron and cobalt, *Phys. Rev. Lett.* **75**, p. 152.

Chen, X., Na, M., Cheon, M., Wang, S., Luo, H., McCombe, B. D., Liu, X., Sasaki, Y., Wojtowicz, T., Furdyna, J. K., Potashnik, S. J. and Schiffer, P. (2002). Above-room-temperature ferromagnetism in GaSb/Mn digital alloys, *Appl. Phys. Lett.* **81**, p. 511.

Cheng, S., Nadgomy, B., Bussmann, K., Carpenter, E., Das, B., Trotter, G., Raphael, M. and Harris, V. (2001). Growth and magnetic properties of single crystal Co_2MnX (X=Si,Ge) Heusler alloys, *IEEE Transactions on Magnetics* **37**, p. 2176.

Clowes, S., Miyoshi, Y., Bugoslavsky, Y., Branford, W., Grigorescu, C., Manea, S., Monnereau, O. and Cohen, L. (2004). Spin polarization of the transport current at the free surface of bulk NiMnSb, *Phys. Rev. B* **69**, p. 214425.

Coey, J. M. D. and Venkatesan, M. (2002). Half-metallic ferromagnetism: Example of CrO_2 (invited), *J. Appl. Phys.* **91**, p. 8345.

Coey, J. M. D., Versluijs, J. J. and Venkatesan, M. (2002). Half-metallic oxide point contacts, *J. Phys. D: Appl. Phys.* **35**, p. 2457.

Continenza, A., Picozzi, S., Geng, W. T. and Freeman, A. J. (2001). Coordination and chemical effects on the structural, electronic, and magnetic properties in Mn pnictides, *Phys. Rev. B* **64**, p. 085204.

Dag, S., Tongay, S., Yildirim, T., Durgun, E., Senger, R., Fong, C. Y. and Ciraci, S. (2005). Half-metallic properties of atomic chains of carbon-transition metal compounds, *Phys. Rev. B* **72**, p. 155444.

Datta, S. and Das, B. (1990). Electronic analog of the electro-optic modulator, *Appl. Phys. Lett.* **56**, p. 665.

Daughton, J. (1992). Magnetoresistive memory technology, *Thin Solid Films* **216**, p. 162.

de Groot, R. A., Mueller, F., Engen, P. and Buschow, K. (1983). New class of materials: Half-metallic ferromagnets, *Phys. Rev. Lett.* **50**, p. 2024.

De Teresa, J. M., Barthélémy, A., Fert, A., Contour, J. P., Lyonnet, R., Montaigne, F., Seneor, P. and Vaurès, A. (1999). Inverse tunnel magnetoresistance in$Co/SrTiO_3/La_{0.7}Sr_{0.3}MnO_3$: New ideas on spin-polarized tunneling, *Phys. Rev. Lett.* **82**, p. 4288.

de Wijs, G. A. and de Groot, R. A. (2001). Towards 100% spin-polarized charge-injection: The half-metallic NiMnSb/CdS interface, *Phys. Rev. B* **64**, p. 020402.

Dedkov, Y. (2004). *Spin-Resolved Photoelectron Spectroscopy of Oxidic Half-Metallic Ferromagnets and Oxide/Ferromagnet Interfaces*, Ph.D. thesis, Technical University of Aachen, Aachen, Germany.

Dedkov, Y. S., Rüdiger, U. and Güntherodt, G. (2002). Evidence for the half-metallic ferromagnetic state of Fe_3O_4 by spin-resolved photoelectron spectroscopy, *Phys. Rev. B* **65**, p. 064417.

Dieny, B., Humbert, P., Speriosu, V. S., Metin, S., Gurney, B. A., Baugart, P. and Lefakis, H. (1992). Giant magnetoresistance of magnetically soft sandwiches: Dependence on temperature and on layer thicknesses, *Phys. Rev. B* **45**, p. 806.

Dowben, P. A. and Skomski, R. (2004). Are half-metallic ferromagnets half metals? (invited), *J. Appl. Phys.* **95**, p. 7453.

Dupree, T. H. (1961). Electron scattering in a crystal lattice, *Ann. Phys. -New York* **15**, p. 63.

Durgun, E., Senger, R. T., Mehrez, H., Dag, S. and Ciraci, S. (2006). Nanospintronic properties of carbon-cobalt atomic chains, *EPL (Europhysics Letters)* **73**, p. 642.

Eerenstein, W., Palstra, T. T. M., Hibma, T. and Celotto, S. (2002). Origin of the increased resistivity in epitaxial Fe_3O_4 films, *Phys. Rev. B* **66**, p. 201101.

Elmers, H., Wurmehl, S., Fecher, G., Jakob, G., Felser, C. and Schönhense, G. (2004). Field dependence of orbital magnetic moments in the Heusler compounds Co_2FeAl and $Co_2Cr_{0.6}Fe_{0.4}Al$, *Appl. Phys. A–Mater. Sci. Process.* **79**, p. 557.

Etgens, V. H., de Camargo, P. C., Eddrief, M., Mattana, R., George, J. M. and Garreau, Y. (2004). Structure of ferromagnetic CrAs epilayers grown on GaAs(001), *Phys. Rev. Lett.* **92**, p. 167205.

Feng, J. S. Y., Pashley, R. D. and Nicolet, M. A. (1975). Magnetoelectric properties of magnetite thin films, *J. Phys. C: Solid State Phys.* **8**, p. 1010.

Foner, S. (1956). Vibrating sample magnetometer, *Rev. Sci. Instrum.* **27**, p. 548.

Fong, C. Y. and Cohen, M. L. (1970). Energy band structure of copper by the empirical pseudopotential method, *Phys. Rev. Lett.* **24**, p. 306.

Fong, C. Y. and Qian, M. C. (2004). New spintronic superlattices composed of half-metallic compounds with zinc-blende structure, *J. Phys.: Condens. Mat.* **16**, p. S5669.

Fong, C. Y., Qian, M. C., Liu, K., Yang, L. H. and Pask, J. E. (2008). Design of spintronic materials with simple structures, *J. Nanosci. Nanotechnol.* **8**, p. 3652.

Fong, C. Y., Qian, M. C., Pask, J. E., Yang, L. H. and Dag, S. (2004). Electronic and magnetic properties of zinc blende half-metal superlattices, *Appl. Phys. Lett.* **84**, p. 239.

Fuji, S., Sugimurat, S., Ishidat, S. and Asano, S. (1990). Hyperfine fields and electronic structures of the Heusler alloys Co_2MnX (X=Al, Ga, Si, Ge, Sn), *J. Phys.: Condens. Mat.* **2**, p. 8583.

Furutani, Y., Nishihara, H., Kanomata, T., Kobayashi, K., Kainuma, R., Ishida, K., Koyama, K., Watanabe, K. and Goto, T. (2009). Field-induced-moment nuclear coupling for [59]Co in a Heusler alloyCo_2TiGa, *J. Phys.: Conf. Ser.* **150**, p. 042037.

Galanakis, I. (2002a). Surface half-metallicity of CrAs in the zinc-blende structure, *Phys. Rev. B* **66**, p. 012406.

Galanakis, I. (2002b). Surface properties of the half- and full-Heusler alloys, *J. Phys.: Condens. Mat.* **14**, p. 6329.

Galanakis, I. (2005). Orbital magnetism in the half-metallic Heusler alloys, *Phys. Rev. B* **71**, p. 012413.

Galanakis, I., Dederichs, P. and Papanikolaou, N. (2002a). Origin and properties of the gap in the half-ferromagnetic Heusler alloys, *Phys. Rev. B* **66**, p. 134428.

Galanakis, I., Dederichs, P. and Papanikolaou, N. (2002b). Slater–Pauling behavior and origin of the half-metallicity of the full-Heusler alloys, *Phys. Rev. B* **66**, p. 174429.

Galanakis, I. and Mavropoulos, P. (2003). Zinc-blende compounds of transition elements with N, P, As, Sb, S, Se, and Te as half-metallic systems, *Phys. Rev. B* **67**, p. 104417.

Galanakis, I., Ostanin, S., Alouani, M., Dreyssé, H. and Wills, J. (2000). *Ab initio* ground state and $L_{2,3}$ X-ray magnetic circular dichroism of Mn-based Heusler alloys, *Phys. Rev. B* **61**, p. 4093.

García, N., Muñoz, M. and Zhao, Y.-W. (1999). Magnetoresistance in excess of 200% in ballistic Ni nanocontacts at room temperature and 100 Oe, *Phys. Rev. Lett.* **82**, p. 2923.

Gellrich, A. and Kessler, J. (1991). Precision measurement of the Sherman asymmetry function for electron scattering from gold, *Phys. Rev. A* **43**, p. 204.

Getzlaff, M., Heidemann, B., Bansmann, J., Westphal, C. and Schonhense, G. (1998). A variable-angle electron spin polarization detection system, *Rev. Sci. Instrum.* **69**, p. 3913.

Giapintzakis, J., Grigorescu, C., Klini, A., Manousaki, A., Zorba, V., Androulakis, J., Viskadourakis, Z. and Fotakis, C. (2002). Low-temperature growth of NiMnSb thin films by pulsed-laser deposition, *Appl. Phys. Lett.* **80**, p. 2716.

Gong, G. Q., Gupta, A., Xiao, G., Qian, W. and Dravid, V. P. (1997). Magnetoresistance and magnetic properties of epitaxial magnetite thin films, *Phys. Rev. B* **56**, p. 5096.

Goodenough, J. B. (1955). Theory of the role of covalence in the perovskite-type manganites [La,M(II)]MnO_3, *Phys. Rev.* **100**, p. 564.

Grave, E. D., Persoons, R. M., Vandenberghe, R. E. and de Bakker, P. M. A. (1993). Mössbauer study of the high-temperature phase of Co-substituted magnetites, $Co_x Fe_{3-x} O_4$. I. $x \leq 0.04$, *Phys. Rev. B* **47**, p. 5881.

Gridin, V. V., Hearne, G. R. and Honig, J. M. (1996). Magnetoresistance extremum at the first-order Verwey transition in magnetite (Fe_3O_4), *Phys. Rev. B* **53**, p. 15518.

Grollier, J., Cros, V., Hamzic, A., George, J. M., Jaffres, H., Fert, A., Faini, G., Youssef, J. B. and Legall, H. (2001). Spin-polarized current induced switching in Co/Cu/Co pillars, *Appl. Phys. Lett.* **78**, p. 3663.

Grossu, G. and Paravocicino, G. P. (2000). *Solid State Physics* (Academic Press, New York).

Gupta, A., Li, X. W., Guha, S. and Xiao, G. (1999). Selective-area and lateral overgrowth of chromium dioxide (CrO_2) films by chemical vapor deposition, *Appl. Phys. Lett.* **75**, p. 2996.

Gupta, A., Li, X. W. and Xiao, G. (2000). Magnetic and transport properties of epitaxial and polycrystalline chromium dioxide thin films (invited), *J. Appl. Phys.* **87**, p. 6073.

Halilov, S. V. and Kulatov, E. T. (1991). Electron and magneto-optical properties of half-metallic ferromagnets and uranium monochalcogenide, *J. Phys.: Condens. Mat.* **3**, p. 6363.

Hamada, N., Sawada, H. and Terakuar, K. (1995). *Spectroscopy of Mott Insulators and Correlated Metal*, Springer Series in Solid-State Sciences 119 (Springer-Verlag, Berlin).

Hamann, D., Schlüter, M. and Chiang, C. (1979). Norm-conserving pseudopotentials, *Phys. Rev. Lett.* **43**, p. 1494.

Hamrle, J., Blomeier, S., Gaier, O., Hillebrands, B., Schneider, H., Jakob, G., Postava, K. and Felser, C. (2007). Huge quadratic magneto-optical Kerr effect and magnetization reversal in the Co_2FeSi Heusler compound, *J. Phys. D: Appl. Phys.* **40**, p. 1563.

Hanssen, K. and Mijnarends, P. (1986). Positron-annihilation study of the half-metallic ferromagnet NiMnSb: Theory, *Phys. Rev. B* **34**, p. 5009.

Hanssen, K., Mijnarends, P., Rabou, L. and Buschow, K. (1990). Positron-annihilation study of the half-metallic ferromagnet NiMnSb: Experiment, *Phys. Rev. B* **42**, p. 1533.

Harris, J. J., Joyce, B. A. and Dobson, P. (1981). Oscillations in the surface structure of Sn-doped GaAs during growth by MBE, *Surf. Sci.* **103**, p. L90.

Hashimoto, M., Herfort, J., Schonherr, H.-P. and Ploog, K. (2005). Epitaxial Heusler alloy $Co_2FeSi/GaAs(001)$ hybrid structures, *Appl. Phys. Lett.* **87**, p. 102506.

Heine, V. (1980). Electronic structure from the point of view of the local atomic environment, in D. Turnbull, H. Ehrenreich and F. Seitz (eds.), *Solid State Physics*, Vol. 35 (Academic Press, New York), p. 1.

Heinz, K. (1995). LEED and DLEED as modern tools for quantitative surface structure determination, *Rep. Prog. Phys.* **58**, p. 637.

Helmholdt, R. B., de Groot, R. A., Mueller, F. M., van Engen, P. G. and Buschow, K. H. J. (1984). Magnetic and crystallographic properties of several $C1_b$ type Heusler compounds, *J. Magn. Magn. Mater.* **43**, p. 249.

Herring, C. (1940). A new method for calculating wave functions in crystals, *Phys. Rev.* **57**, p. 1169.

Heusler, F. (1903). über magnetische manganlegierungen, *Verh. Dtsch. Phys. Ges.* **5**, p. 219.

Hirohata, A., Kikuchi, M., Tezuka, N., Inomata, K., Claydon, J., Xu, Y. and van der Laan, G. (2006). Heusler alloy/semiconductor hybrid structures, *Current Opinion in Solid State and Materials Science* **10**, p. 93.

Hirohata, A., Kurebayashi, H., Okamura, S., Kikuchi, M., Masaki, T., Nozaki, T., Tezuka, N. and Inomata, K. (2005). Structural and magnetic properties of epitaxial $L2_1$–structured $Co_2(Cr,Fe)Al$ films grown on GaAs(001) substrates, *J. Appl. Phys.* **97**, p. 103714.

Hohenberg, P. and Kohn, W. (1964). Inhomogeneous electron gas, *Phys. Rev.* **136**, p. B864.

Hordequin, C., Pierre, J. and Currat, R. (1996). Magnetic excitations in the half-metallic NiMnSb ferromagnet: From Heisenberg-type to itinerant behaviour, *J. Magn. Magn. Mater.* **162**, p. 75.

Hordequin, C., Ristoiu, D., Ranno, L. and Pierre, J. (2000). On the cross-over from half-metal to normal ferromagnet in NiMnSb, *Euro. Phys. J. B* **16**, p. 287.

Hörmandinger, G., Weinberger, P., Marksteiner, P. and Redinger, J. (1988). Theoretical calculations of core-core-valence Auger spectra: Applications to the $L_3M_{2,3}V$ transitions of Ti in nonstoichiometric Ti-C, Ti-N, and Ti-O, *Phys. Rev. B* **38**, p. 1040.

Hu, G. and Suzuki, Y. (2002). Negative spin polarization of Fe_3O_4 in magnetite/manganite-based junctions, *Phys. Rev. Lett.* **89**, p. 276601.

Huang, D. J., Chang, C. F., Chen, J., Tjeng, L. H., Rata, A. D., Wu, W. P., Chung, S. C., Lin, H. J., Hibma, T. and Chen, C. T. (2002). Spin-resolved photoemission studies of epitaxial Fe_3O_4 (100) thin films, *J. Magn. Magn. Mater.* **239**, p. 261.

Hubbard, J. (1963). Electron correlations in narrow energy bands, *Proc. R. Soc. Lond. A* **276**, p. 238.

Husmann, A. and Singh, L. (2006). Temperature dependence of the anomalous Hall conductivity in the Heusler alloy Co_2CrAl, *Phys. Rev. B* **73**, p. 172417.

Hwang, H. Y. and Cheong, S.-W. (1997). Enhanced intergrain tunneling magnetoresistance in half-metallic CrO_2 films, *Science* **278**, p. 1607.

Hwang, H. Y., Cheong, S.-W., Ong, N. P. and Batlogg, B. (1996). Spin-polarized intergrain tunneling in $La_{2/3}Sr_{1/3}MnO_3$, *Phys. Rev. Lett.* **77**, p. 2041.

Iizumi, M., Koetzle, T. F., Shirane, G., Chikazumi, S., Matsui, M. and Todo, S. (1982). Structure of magnetite Fe_3O_4 below the Verwey transition temperature, *Acta Crystall. B* **38**, p. 2121.

Inomata, K., Okamura, S., Miyazaki, A., Kikuchi, M., Tezuka, N., Wojcik, M. and Jedryka, E. (2006). Structural and magnetic properties and tunnel magnetoresistance for $Co_2(Cr,Fe)Al$ and Co_2FeSi full-Heusler alloys, *J. Phys. D: Appl. Phys.* **39**, p. 816.

Ishida, S., Akazawa, S., Kubo, Y. and Ishida, J. (1982). Band theory of Co_2MnSn, Co_2TiSn and Co_2TiAl, *J. Phys. F: Met. Phys.* **12**, p. 1111.

Ishida, S., Masaki, T., Fujii, S. and Asano, S. (1998). Theoretical search for half-metallic films of Co_2MnZ (Z = Si, Ge), *Physica B: Condensed Matter* **245**, p. 1.

Ivanov, P. G., Watts, S. M. and Lind, D. M. (2001). Epitaxial growth of CrO_2 thin films by chemical-vapor deposition from a Cr_8O_{21} precursor, *J. Appl. Phys.* **89**, p. 1035.

Janak, J., Moruzzi, V. and Williams, A. (1975). Ground-state thermomechanical properties of some cubic elements in the local-density formalism, *Phys. Rev. B* **12**, p. 1257.

Jeng, H.-T., Guo, G. Y. and Huang, D. J. (2004). Charge-orbital ordering and Verwey transition in magnetite, *Phys. Rev. Lett.* **93**, p. 156403.

Jenkins, S. J. (2004). Ternary half-metallics and related binary compounds: Stoichiometry, surface states, and spin, *Phys. Rev. B* **70**, p. 245401.

Ji, Y., Strijkers, G., Yang, F., Chien, C., Byers, J., Anguelouch, A., Xiao, G. and Gupta, A. (2001). Determination of the spin polarization of half-metallic CrO_2 by point contact Andreev reflection, *Phys. Rev. Lett.* **86**, p. 5585.

Jonker, G. H. and Santen, J. H. V. (1950). Ferromagnetic compounds of manganese with perovskite structure, *Physica* **16**, p. 337.

Kabani, R., Terada, M., Roshko, A. and Moodera, J. S. (1990). Magnetic properties of NiMnSb films, *J. Appl. Phys.* **67**, p. 4898.

Kallmayer, M., Schneider, H., Jakob, G., Elmers, H. J., Balke, B. and Cramm, S. (2007). Interface magnetization of ultrathin epitaxial $Co_2FeSi(110)/Al_2O_3$ films, *J. Phys. D: Appl. Phys.* **40**, p. 1552.

Kämmerer, S., Heitmann, S., Meyners, D., Sudfeld, D., Thomas, A., Hütten, A. and Reiss, G. (2003). Room-temperature preparation and magnetic behavior of Co_2MnSi thin films, *J. Appl. Phys.* **93**, p. 7945.

Kämmerer, S., Thomas, A., Hütten, A. and Reiss, G. (2004). Co_2MnSi Heusler alloy as magnetic electrodes in magnetic tunnel junctions, *Appl. Phys. Lett.* **85**, p. 79.

Kämper, K. P., Schmitt, W., Güntherodt, G., Gambino, R. J. and Ruf, R. (1987). CrO_2 — a new half-metallic ferromagnet? *Phys. Rev. Lett.* **59**, p. 2788.

Kandpal, H., Fecher, G., Felser, C. and Schönhense, G. (2006). Correlation in the transition-metal-based Heusler compounds Co_2MnSi and Co_2FeSi, *Phys. Rev. B* **73**, p. 094422.

Kang, J.-S., Park, J.-G., Olson, C. G., Younq, S. J. and Min, B. I. (1995). Valence band and Sb 4d core level photoemission of the XMnSb-type Heusler compounds (X=Pt,Pd,Ni), *J. Phys.: Condens. Mat.* **7**, p. 3789.

Karthik, S., Rajanikanth, A., Nakatani, T., Gercsi, Z., Takahashi, Y., Furubayashi, T., Inomata, K. and Hono, K. (2007). Effect of Cr substitution for Fe on the spin polarization of $Co_2Cr_xFe_{1-x}Si$ Heusler alloys, *J. Appl. Phys.* **102**, p. 043903.

Katsnelson, M. I., Irkhin, V. Y., Chioncel, L., Lichtenstein, A. I. and de Groot, R. A. (2008). Half-metallic ferromagnets: From band structure to many-body effects, *Rev. Mod. Phys.* **80**, p. 315.

Kautzky, M. C., Mancoff, F. B., Bobo, J. F., Johnson, P. R., White, R. L. and Clemens, B. M. (1997). Investigation of possible giant magnetoresistance limiting mechanisms in epitaxial PtMnSb thin films, *J. Appl. Phys.* **81**, p. 4026.

Kawakami, R. K., Johnston-Halperin, E., Chen, L. F., Hanson, M., Guebels, N., Speck, J. S., Gossard, A. C. and Awschalom, D. D. (2000). (Ga,Mn)As as a digital ferromagnetic heterostructure, *Appl. Phys. Lett.* **77**, p. 2379.

Kawano, H., Kajimoto, R., Kubota, M. and Yoshizawa, H. (1996a). Canted antiferromagnetism in an insulating lightly doped $La_{1-x}Sr_xMnO_3$ with $x \leq 0.17$, *Phys. Rev. B* **53**, p. 2202.

Kawano, H., Kajimoto, R., Kubota, M. and Yoshizawa, H. (1996b). Ferromagnetism-induced reentrant structural transition and phase diagram of the lightly doped insulator $La_{1-x}Sr_xMnO_3$ with $x \leq 0.17$, *Phys. Rev. B* **53**, p. R14709.

Kelekar, R. and Clemens, B. (2004). Epitaxial growth of the Heusler alloy $Co_2Cr_{1-x}Fe_xAl$, *J. Appl. Phys.* **96**, p. 540.

Kessler, J. (1985). *Polarized Electrons* (Springer-Verlag, Berlin and New York).

Khoi, L. D., Veillet, P. and Campbell, I. A. (1978). Hyperfine fields and magnetic interactions in Heusler alloys, *J. Phys. F: Met. Phys.* **8**, p. 1811.

Kim, D. J. (1999). *New Perspectives in Magnetism of Metals* (Plenum Publications, New York).

Kim, H.-J., Park, J.-H. and Vescovo, E. (2000a). $Fe_3O_4(111)/Fe(110)$ magnetic bilayer: Electronic and magnetic properties at the surface and interface, *Phys. Rev. B* **61**, p. 15288.

Kim, H.-J., Park, J.-H. and Vescovo, E. (2000b). Oxidation of the Fe(110) surface: An $Fe_3O_4(111)$/Fe(110) bilayer, *Phys. Rev. B* **61**, p. 15284.

Kimura, A., Suga, S., Shishidou, T., Imada, S., Muro, T., Park, S., Miyahara, T., Kaneko, T. and Kanomata, T. (1997). Magnetic circular dichroism in the soft-x-ray absorption spectra of Mn-based magnetic intermetallic compounds, *Phys. Rev. B* **56**, p. 6021.

Kirillova, M., Makhnev, A., Shreder, E., Dyakina, V. and Gorina, N. (1995). Interband optical absorption and plasma effects in half-metallic XMnY ferromagnets, *physica status solidi (b)* **187**, p. 231.

Kittel, C. (2004). *Introduction to Solid State Physics, 8th edition* (Wiley & Sons, New York).

Kleinman, L. and Bylander, D. (1982). Efficacious form for model pseudopotentials, *Phys. Rev. Lett.* **48**, p. 1425.

Kodama, K., Furubayashi, T., Sukegawa, H., Nanktani, T. M., Inomata, K. and Hono, K. (2009). Current-perpendicular-to-plane giant magnetoresistance of a spin valve using Co_2MnSi Heulster alloy electrodes, *J. Appl. Phys.* **105**, p. 07E905.

Koelling, D. D. and Arbman, G. O. (1975). Use of energy derivative of the radial solution in an augmented plane wave method: application to copper, *J. Phys. F: Met. Phys.* **5**, p. 2041.

Kohn, A., Lazarov, V., Singh, L., Barber, Z. and Petford-Long, A. (2007). The structure of sputter-deposited Co_2MnSi thin films deposited on GaAs(001), *J. Appl. Phys.* **101**, p. 023915.

Kohn, W. and Sham, L. (1965). Self-consistent equations including exchange and correlation effects, *Phys. Rev.* **140**, p. A1133.

Kolev, H., Rangelov, G., Braun, J. and Donath, M. (2005). Reduced surface magnetization of NiMnSb(001), *Phys. Rev. B* **72**, p. 104415.

Komesu, T., Borca, C. N., Jeong, H.-K., Dowben, P. A., Ristoiu, D., Nozières, J. P., Stadler, S. and Idzerda, Y. U. (2000). The polarization of Sb overlayers on NiMnSb(100), *Phys. Lett. A* **273**, p. 245.

Korotin, M. A., Anisimov, V. I., Khomskii, D. I. and Sawatzky, G. A. (1998). CrO_2: A self-doped double exchange ferromagnet, *Phys. Rev. Lett.* **80**, p. 4305.

Kouvel, J. S. and Rodbell, D. S. (1967). Magnetic critical-point behavior of CrO_2, *J. Appl. Phys.* **38**, p. 979.

Kramers, H. A. (1934). L'interaction entre les atomes magnètogènes dans un cristal paramagnètique, *Physica* **1**, p. 182.

Kresse, G. and Joubert, D. (1999). From ultrasoft pseudopotentials to the projector augmented-wave method, *Phys. Rev. B* **59**, p. 1758.

Kübler, J. (1984). First principle theory of metallic magnetism, *Physica B+C* **127**, p. 257.

Kubo, K. and Ohata, N. (1972). A quantum theory of double exchange. I, *J. Phys. Soc. Jpn.* **33**, p. 21.

Kulatov, E. and Mazin, I. I. (1990). Extended stoner factor calculations for the half-metallic ferromagnets NiMnSb and CrO_2, *J. Phys.: Condens. Mat.* **2**, p. 343.

Kuneš, J., Novák, P., Oppeneer, P. M., König, C., Fraune, M., Rüdiger, U., Güntherodt, G. and Ambrosch-Draxl, C. (2002). Electronic structure of CrO_2 as deduced from its magneto-optical Kerr spectra, *Phys. Rev. B* **65**, p. 165105.

Langreth, D. C. and Mehl, M. J. (1983). Beyond the local-density approximation in calculations of ground-state electronic properties, *Phys. Rev. B* **28**, p. 1809.

Langreth, D. C. and Perdew, J. P. (1980). Theory of nonuniform electronic systems. I. analysis of the gradient approximation and a generalization that works, *Phys. Rev. B* **21**, p. 5469.

Larson, P., Mahanti, S. D. and Kanatzidis, M. G. (2000). Structural stability of Ni-containing half-Heusler compounds, *Phys. Rev. B* **62**, p. 12754.

Lee, M. J. G. and Falicov, L. M. (1968). The de Haas–van Alphen effect and the Fermi surface of potassium, *Proc. R. Soc. Lond. A* **304**, p. 319.

Lee, W.-L., Watauchi, S., Miller, V. L., Cava, R. J. and Ong, N. P. (2004). Dissipationless anomalous Hall current in the ferromagnetic spinel $CuCr_2Se_{4-x}Br_x$, *Science* **303**, p. 1647.

Li, X. W., Gupta, A., McGuire, T. R., Duncombe, P. R. and Xiao, G. (1999). Magnetoresistance and Hall effect of chromium dioxide epitaxial thin films, *J. Appl. Phys.* **85**, p. 5585.

Liechtenstein, A. I., Anisimov, V. I. and Zaanen, J. (1995). Density-functional theory and strong interactions: Orbital ordering in Mott-Hubbard insulators, *Phys. Rev. B* **52**, p. R5467.

Liechtenstein, A. I., Katsnelson, M. I., Antropov, V. P. and Gubanov, V. A. (1987). Local spin density functional approach to the theory of exchange interactions in ferromagnetic metals and alloys, *J. Magn. Magn. Mater.* **67**, p. 65.

Liu, B.-G. (2003). Robust half-metallic ferromagnetism in zinc-blende CrSb, *Phys. Rev. B* **67**, p. 172411.

Loos, J. and Novák, P. (2002). Double exchange and superexchange in a ferrimagnetic half-metal, *Phys. Rev. B* **66**, p. 132403.

Lu, Y., Li, X. W., Gong, G. Q., Xiao, G., Gupta, A., Lecoeur, P., Sun, J. Z., Wang, Y. Y. and Dravid, V. P. (1996). Large magnetotunneling effect at low magnetic fields in micrometer-scale epitaxial $La_{0.67}Sr_{0.33}MnO_3$ tunnel junctions, *Phys. Rev. B* **54**, p. R8357.

Luo, H., McCombe, B. D., Na, M. H., Mooney, K., Lehmann, F., Chen, X., Cheon, M., Wang, S. M., Sasaki, Y., Liu, X. and Furdyna, J. K. (2002). Transport and magnetic properties of ferromagnetic GaAs/Mn digital alloys, *Physica E: Low-dimensional Systems and Nanostructures* **12**, p. 366.

Ma, C., Yang, Z. and Picozzi, S. (2006). *Ab initio* electronic and magnetic structure in $La_{0.66}Sr_{0.33}MnO_3$: Strain and correlation effects, *J. Phys.: Condens. Mat.* **18**, p. 7717.

Mackintosh, A. R. and Andersen, O. K. (1980). The electronic structure of transition metals, in M. Springford (ed.), *Electrons at the Fermi Surface* (Cambridge University Press, Cambridge), p. 149.

Malozemoff, A., Williams, A. and Moruzzi, V. (1984). "Band-gap theory" of strong ferromagnetism: Application to concentrated crystalline and amorphous Fe- and Co-metalloid alloys, *Phys. Rev. B* **29**, p. 1620.

Mancoff, F., Bobo, J., Richter, O., Bessho, K., Johnson, P., Sinclair, R., Nix, W., White, R. and Clemens, B. (1999). Growth and characterization of epitaxial NiMnSb/PtMnSb Cl_b Heusler alloy superlattices, *J. Mater. Res.* **14**, p. 1560.

Marcus, P. M. and Moruzzi, V. L. (1988). Stoner model of ferromagnetism and total-energy band theory, *Phys. Rev. B* **38**, p. 6949.

Martin, R. M. (2004). *Electronic Structure: Basic Theory and Practical Methods* (Cambridge University Press, Cambridge).

Matsubara, K., Anno, H., Kaneko, H. and Imai, Y. (1999). Electrical properties of half-metallic PtMnSb-based Heusler alloys, in *18th International Conference on Thermoelectrics (1999)*, p. 60.

Mazin, I. I., Singh, D. J. and Ambrosch-Draxl, C. (1999). Transport, optical, and electronic properties of the half-metal CrO_2, *Phys. Rev. B* **59**, p. 411.

McCormack, M., Jin, S., Tiefel, T. H., Fleming, R. M., Phillips, J. M. and Ramesh, R. (1994). Very large magnetoresistance in perovskite-like La–Ca–Mn–O thin films, *Appl. Phys. Lett.* **64**, p. 3045.

McQueeney, R. J., Yethiraj, M., Montfrooij, W., Gardner, J. S., Metcalf, P. and Honig, J. M. (2006). Investigation of the presence of charge order in magnetite by measurement of the spin wave spectrum, *Phys. Rev. B* **73**, p. 174409.

Meservey, R. and Tedrow, P. M. (1994). Spin-polarized electron tunneling, *Physics Reports* **238**, p. 173.

Metalidis, G. and Bruno, P. (2006). Inelastic scattering effects and the Hall resistance in a four-probe ring, *Phys. Rev. B* **73**, p. 113308.

Miyamoto, K., Kimura, A., Iori, K., Sakamoto, K., Xie, T., Moko, T., Qiao, S., Taniguchi, M. and Tsuchiya, K. (2004). Element-resolved magnetic moments of Heusler-type ferromagnetic ternary alloy Co_2MnGe, *J. Phys.: Condens. Mat.* **16**, p. S5797.

Mizuguchi, M., Akinaga, H., Manago, T., Ono, K., Oshima, M., Shirai, M., Yuri, M., Lin, H. J., Hsieh, H. H. and Chen, C. T. (2002). Epitaxial growth of zinc-blende CrAs/GaAs multilayer, *J. Appl. Phys.* **91**, p. 7917.

Moore, G. E. (1965). Cramming more components onto integrated circuits, *Electronics* **38**, p. 114.

Moos, R., Menesklou, W. and Härdtl, K. H. (1995). Hall mobility of undoped n-type conducting strontium titanate single crystals between 19 K and 1373 K, *Appl. Phys. A* **61**, p. 389.

Mukhamedzhanov, E. K., Bocchi, C., Franchi, S., Baraldi, A., Magnanini, R. and Nasi, L. (2000). High-resolution x-ray diffraction, x-ray standing-wave, and transmission electron microscopy study of Sb-based single-quantum-well structures, *J. Appl. Phys.* **87**, p. 4234.

Nazarenko, E., Lorenzo, J. E., Joly, Y., Hodeau, J. L., Mannix, D. and Marin, C. (2006). Resonant X-Ray diffraction studies on the charge ordering in magnetite, *Phys. Rev. Lett.* **97**, p. 056403.

Nèel, L. (1948). Magnetic properties of ferrites: Ferrimagnetism and antiferro-magnetism, *Ann. Phys.-Paris* **3**, p. 137.

Niculescu, V., Burch, T., Raj, K. and Budnick, J. (1977). Properties of Heusler-type materials Fe_2TSi and $FeCo_2Si$, *J. Magn. Magn. Mater.* **5**, p. 60.

Norby, P., Christensen, A. N., Fjellvåag, H. and Nielsen, M. (1991). The crystal structure of Cr_8O_{21} determined from powder diffraction data: Thermal transformation and magnetic properties of a chromium-chromate-tetrachromate, *J. Solid State Chem.* **94**, p. 281.

Oguchi, T., Terakura, K. and Williams, A. R. (1984). Transition-metal monoxides: Itinerant versus localized picture of superexchange, *J. Appl. Phys.* **55**, p. 2318.

Ohashi, S., Lippmaa, M., Nakagawa, N., Nagasawa, H., Koinuma, H. and Kawasaki, M. (1999). Compact laser molecular beam epitaxy system using laser heating of substrate for oxide film growth, *Rev. Sci. Instrum.* **70**, p. 178.

Okabayashi, J., Mizuguchi, M., Ono, K., Oshima, M., Fujimori, A., Kuramochi, H. and Akinaga, H. (2004). Density-dependent electronic structure of zinc-blende-type MnAs dots on GaAs(001) studied by *in situ* photoemission spectroscopy, *Phys. Rev. B* **70**, p. 233305.

Okimoto, Y., Katsufuji, T., Ishikawa, T., Urushibara, A., Arima, T. and Tokura, Y. (1995). Anomalous variation of optical spectra with spin polarization in double-exchange ferromagnet: $La_{1-x}Sr_xMnO_3$, *Phys. Rev. Lett.* **75**, p. 109.

Ono, K., Okabayashi, J., Mizuguchi, M., Oshima, M., Fujimori, A. and Akinaga, H. (2002). Fabrication, magnetic properties, and electronic structures of nanoscale zinc-blende MnAs dots (invited), *J. Appl. Phys.* **91**, p. 8088.

Orgassa, D., Fujiwara, H., Schulthess, T. C. and Butler, W. H. (1999). First-principles calculation of the effect of atomic disorder on the electronic structure of the half-metallic ferromagnet NiMnSb, *Phys. Rev. B* **60**, p. 13237.

Orgassa, D., Fujiwara, H., Schulthess, T. C. and Butler, W. H. (2000). Disorder dependence of the magnetic moment of the half-metallic ferromagnet NiMnSb from first principles, *J. Appl. Phys.* **87**, p. 5870.

Oswald, A., Zeller, R., Braspenning, P. J. and Dederichs, P. H. (1985). Interaction of magnetic impurities in Cu and Ag, *J. Phys. F: Met. Phys.* **15**, p. 193.

Otto, M. J., van Woerden, R. A. M., van der Valk, P. J., Wijngaard, J., van Bruggen, C. F., Haas, C. and Buschow, K. H. J. (1989). Half-metallic ferromagnets. I. structure and magnetic properties of NiMnSb and related inter-metallic compounds, *J. Phys.: Condens. Mat.* **1**, p. 2341.

Paiva-Santos, C. O., Marques, R. F. C., Jafelicci, J. M. and Varanda, L. C. (2002). X-ray powder data and bond valence of $La_{0.65}Sr_{0.35}MnO_3$ after Rietveld refinement, *Powder Diffraction* **17**, p. 149.

Pajda, M., Kudrnovský, J., Turek, I., Drchal, V. and Bruno, P. (2001). *Ab initio* calculations of exchange interactions, spin-wave stiffness constants, and Curie temperatures of Fe, Co, and Ni, *Phys. Rev. B* **64**, p. 174402.

Park, J.-H., Vescovo, E., Kim, H.-J., Kwon, C., Ramesh, R. and Venkatesan, T. (1998a). Direct evidence for a half-metallic ferromagnet, *Nature* **392**, p. 794.

Park, J.-H., Vescovo, E., Kim, H.-J., Kwon, C., Ramesh, R. and Venkatesan, T. (1998b). Magnetic properties at surface boundary of a half-metallic ferromagnet $La_{0.7}Sr_{0.3}MnO_3$, *Phys. Rev. Lett.* **81**, p. 1953.

Parkin, S. S. P., Bhadra, R. and Roche, K. P. (1991). Oscillatory magnetic exchange coupling through thin copper layers, *Phys. Rev. Lett.* **66**, p. 2152.

Pask, J. E., Yang, L. H., Fong, C. Y., Pickett, W. E. and Dag, S. (2003). Six low-strain zinc-blende half metals: An *ab initio* investigation, *Phys. Rev. B* **67**, p. 224420.

Peierls, R. E. (1955). *Quantum Theory of Solids* (Oxford University Press, Oxford).

Perdew, J. P., Burke, K. and Ernzerhof, M. (1996). Generalized gradient approximation made simple, *Phys. Rev. Lett.* **77**, p. 3865.

Perdew, J. P., Burke, K. and Ernzerhof, M. (1997). Errata: Generalized gradient approximation made simple [Phys. Rev. Lett. 77, 3865 (1996)], *Phys. Rev. Lett.* **78**, p. 1396.

Perdew, J. P., Chevary, J. A., Vosko, S. H., Jackson, K. A., Pederson, M. R., Singh, D. J. and Fiolhais, C. (1992). Atoms, molecules, solids, and surfaces: Applications of the generalized gradient approximation for exchange and correlation, *Phys. Rev. B* **46**, p. 6671.

Perdew, J. P. and Yue, W. (1986). Accurate and simple density functional for the electronic exchange energy: Generalized gradient approximation, *Phys. Rev. B* **33**, p. 8800.

Perdew, J. P. and Zunger, A. (1981). Self-interaction correction to density-functional approximations for many-electron systems, *Phys. Rev. B* **23**, p. 5048.

Pesavento, P. V., Chesterfield, R. J., Newman, C. R. and Frisbie, C. D. (2004). Gated four-probe measurements on pentacene thin-film transistors: Contact resistance as a function of gate voltage and temperature, *J. Appl. Phys.* **96**, p. 7312.

Phillips, J. C. and Kleinman, L. (1959). New method for calculating wave functions in crystals and molecules, *Phys. Rev.* **116**, p. 287.

Picozzi, S., Continenza, A. and Freeman, A. J. (2002). Co_2Mn_X (X=Si, Ge, Sn) Heusler compounds: An *ab initio* study of their structural, electronic, and magnetic properties at zero and elevated pressure, *Phys. Rev. B* **66**, p. 094421.

Picozzi, S., Continenza, A. and Freeman, A. J. (2004). Role of structural defects on the half-metallic character of Co_2MnGe and Co_2MnSi Heusler alloys, *Phys. Rev. B* **69**, p. 094423.

Podloucky, R., Zeller, R. and Dederichs, P. H. (1980). Electronic structure of magnetic impurities calculated from first principles, *Phys. Rev. B* **22**, p. 5777.

Porta, P., Marezio, M., Remeika, J. P. and Dernier, P. D. (1972). Chromium dioxide: High pressure synthesis and bond lengths, *Materials Research Bulletin* **7**, p. 157.

Postava, K., Hrabovsky, D., Pistora, J., Fert, A. R., Visnovsky, S. and Yamaguchi, T. (2002). Anisotropy of quadratic magneto-optic effects in reflection, *J. Appl. Phys.* **91**, p. 7293.

Poulsen, U. K., Kollar, J. and Andersen, O. K. (1976). Magnetic and cohesive properties from canonical bands (for transition metals), *J. Phys. F: Met. Phys.* **6**, p. L241.

Qian, M. C., Fong, C. Y., Liu, K., Pickett, W. E., Pask, J. E. and Yang, L. H. (2006a). Half-metallic digital ferromagnetic heterostructure composed of a δ-doped layer of Mn in Si, *Phys. Rev. Lett.* **96**, p. 027211.

Qian, M. C., Fong, C. Y. and Pickett, W. E. (2006b). Enhancement of ferromagnetic coupling in Mn/GaAs digital ferromagnetic heterostructure by free-hole injection, *J. Appl. Phys.* **99**, p. 08D517.

Qian, M. C., Fong, C. Y., Pickett, W. E., Pask, J. E., Yang, L. H. and Dag, S. (2005). Spin-polarized ballistic transport in a thin superlattice of zinc blende half-metallic compounds, *Phys. Rev. B* **71**, p. 012414.

Qian, M. C., Fong, C. Y., Pickett, W. E. and Wang, H.-Y. (2004a). An *ab initio* investigation on the zinc-blende MnAs nanocrystallite, *J. Appl. Phys.* **95**, p. 7459.

Qian, M. C., Fong, C. Y. and Yang, L. H. (2004b). Coexistence of localized magnetic moment and opposite-spin itinerant electrons in MnC, *Phys. Rev. B* **70**, p. 052404.

Ranno, L., Barry, A. and Coey, J. M. D. (1997). Production and magnetotransport properties of CrO_2 films, *J. Appl. Phys.* **81**, p. 5774.

Raphael, M., Ravel, B., Huang, Q., Willard, M., Cheng, S., Das, B., Stroud, R., Bussmann, K., Claassen, J. and Harris, V. (2002). Presence of antisite disorder and its characterization in the predicted half-metal Co_2MnSi, *Phys. Rev. B* **66**, p. 104429.

Raue, R., Hopster, H. and Kisker, E. (1984). High-resolution spectrometer for spin-polarized electron spectroscopies of ferromagnetic materials, *Rev. Sci. Instrum.* **55**, p. 383.

Reisinger, D., Schonecke, M., Brenninger, T., Opel, M., Erb, A., Alff, L. and Gross, R. (2003). Epitaxy of Fe_3O_4 on Si(001) by pulsed laser deposition using a TiN/MgO buffer layer, *J. Appl. Phys.* **94**, p. 1857.

Rietveld, H. M. (1969). A profile refinement method for nuclear and magnetic structures, *J. Appl. Cryst.* **2**, p. 65.

Ristoiu, D., Nozires, J. P., Borca, C. N., Borca, B. and Dowben, P. A. (2000a). Manganese surface segregation in NiMnSb, *Appl. Phys. Lett.* **76**, p. 2349.

Ristoiu, D., Nozires, J. P., Borca, C. N., Komesu, T., Jeong, H.-K. and Dowben, P. A. (2000b). The surface composition and spin polarization of NiMnSb epitaxial thin films, *EPL (Europhysics Letters)* **49**, p. 624.

Ritchie, L., Xiao, G., Ji, Y., Chen, T., Chien, C., Zhang, M., Chen, J., Liu, Z., Wu, G. and Zhang, X. (2003). Magnetic, structural, and transport properties of the Heusler alloys Co_2MnSi and $NiMnSb$, *Phys. Rev. B* **68**, p. 104430.

Robinson, I. K. (1991). *Handbook on Synchrotron Radiation*, Vol. 3 (Elsevier, Amsterdam).

Ross, C. (2001). Patterned magnetic recording media, *Ann. Rev. Mater. Res.* **31**, p. 203.

Rusz, J., Bergqvist, L., Kudrnovský, J. and Turek, I. (2006). Exchange interactions and Curie temperatures in $Ni_{2-x}MnSb$ alloys: First-principles study, *Phys. Rev. B* **73**, p. 214412.

Sacchi, M., Spezzani, C., Carpentiero, A., Prasciolu, M., Delaunay, R., Luning, J. and Polack, F. (2007). Experimental setup for lensless imaging via soft x-ray resonant scattering, *Rev. Sci. Instrum.* **78**, p. 043702.

Sakuraba, Y., Hattori, M., Oogane, M., Ando, Y., Kato, H., Sakuma, A., Miyazaki, T. and Kubota, H. (2006). Giant tunneling magnetoresistance in $Co_2MnSi/Al-O/Co_2MnSi$ magnetic tunnel junctions, *Appl. Phys. Lett.* **88**, p. 192508.

Salamon, M. B. and Jaime, M. (2001). The physics of manganites: Structure and transport, *Rev. Mod. Phys.* **73**, p. 583.

Sánchez-Portal, D., Ordejón, P., Artacho, E. and Soler, J. M. (1997). Density-functional method for very large systems with LCAO basis sets, *Int. J. Quant. Chem.* **65**, p. 453.

Sandratskii, L. M. (1998). Noncollinear magnetism in itinerant-electron systems: theory and applications, *Adv. Phys.* **47**, p. 91.

Sanvito, S. and Hill, N. (2001). *Ab initio* transport theory for digital ferromagnetic heterostructures, *Phys. Rev. Lett.* **87**, p. 267202.

Şaşıoğlu, E., Galanakis, I., Sandratskii, L. M. and Bruno, P. (2005a). Stability of ferromagnetism in the half-metallic pnictides and similar compounds: a first-principles study, *J. Phys.: Condens. Mat.* **17**, p. 3915.

Şaşıoğlu, E., Sandratskii, L. M., Bruno, P. and Galanakis, I. (2005b). Exchange interactions and temperature dependence of magnetization in half-metallic Heusler alloys, *Phys. Rev. B* **72**, p. 184415.

Satpathy, S. and Vukajlović, Z. S. P. F. R. (1996). Electronic structure of the perovskite oxides: $La_{1-x}Ca_xMnO_3$, *Phys. Rev. Lett.* **76**, p. 960.

Savtchenko, L., Engel, B., Rizzo, N., Deherrera, M. and Janesky, J. (2003). Method of writing to scalable magnetoresistance random access memory element, US Patent, 6,545,906 B1.

Schlomka, J.-P., Tolan, M. and Press, W. (2000). *In situ* growth study of $NiMnSb$ films on $MgO(001)$ and $Si(001)$, *Appl. Phys. Lett.* **76**, p. 2005.

Schlottmann, P. (2003). Double-exchange mechanism for CrO_2, *Phys. Rev. B* **67**, p. 174419.

Schmalhorst, J., Kämmerer, S., Sacher, M., Reiss, G., Hütten, A. and Scholl, A. (2004). Interface structure and magnetism of magnetic tunnel junctions with a Co_2MnSi electrode, *Phys. Rev. B* **70**, p. 024426.

Schneider, H., Herbort, C., Jakob, G., Adrian, H., Wurmehl, S. and Felser, C. (2007). Structural, magnetic and transport properties of Co_2FeSi Heusler films, *J. Phys. D: Appl. Phys.* **40**, p. 1548.

Schneider, H., Jakob, G., Kallmayer, M., Elmers, H., Cinchetti, M., Balke, B., Wurmehl, S., Felser, C., Aeschlimann, M. and Adrian, H. (2006). Epitaxial film growth and magnetic properties of Co_2FeSi, *Phys. Rev. B* **74**, p. 174426.

Schwarz, K. (1986). CrO_2 predicted as a half-metallic ferromagnet, *J. Phys. F: Met. Phys.* **16**, p. L211.

Shen, J., Gai, Z. and Kirschner, J. (2004). Growth and magnetism of metallic thin films and multilayers by pulsed-laser deposition, *Surf. Sci. Rep.* **52**, p. 163.

Shepherd, J. P. and Sandberg, C. J. (1984). Measurement and control of oxygen fugacity during annealing, *Rev. Sci. Instrum.* **55**, p. 1696.

Shima, M., Tepper, T. and Ross, C. A. (2002). Magnetic properties of chromium oxide and iron oxide films produced by pulsed laser deposition, *J. Appl. Phys.* **91**, p. 7920.

Shirai, M. (2004). The computational design of zinc-blende half-metals and their nanostructures, *J. Phys.: Condens. Mat.* **16**, p. S5525.

Shull, C. G., Wollan, E. O. and Koehler, W. C. (1951). Neutron scattering and polarization by ferromagnetic materials, *Phys. Rev.* **84**, p. 912.

Singh, D. J. and Nordström, L. (2006). *Pseudopotentials and the LAPW method* (Springer, New York).

Singh, D. J. and Pickett, W. E. (1998). Pseudogaps, Jahn-Teller distortions, and magnetic order in manganite perovskites, *Phys. Rev. B* **57**, p. 88.

Singh, L. J., Barber, Z. H., Kohn, A., Petford-Long, A. K., Miyoshi, Y., Bugoslavsky, Y. and Cohen, L. F. (2006). Interface effects in highly oriented films of the Heusler alloy Co_2MnSi on GaAs(001), *J. Appl. Phys.* **99**, p. 013904.

Singh, L. J., Barber, Z. H., Miyoshi, Y., Branford, W. R. and Cohen, L. F. (2004a). Structural and transport studies of stoichiometric and off-stoichiometric thin films of the full Heusler alloy Co_2MnSi, *J. Appl. Phys.* **95**, p. 7231.

Singh, L. J., Barber, Z. H., Miyoshi, Y., Bugoslavsky, Y., Branford, W. R. and Cohen, L. F. (2004b). Structural, magnetic, and transport properties of thin films of the Heusler alloy Co_2MnSi, *Appl. Phys. Lett.* **84**, p. 2367.

Skriver, H. L. (1983). *The LMTO Method* (Springer-Verlag, Berlin).

Slater, J. C. (1951). Magnetic effects and the Hartree–Fock equation, *Phys. Rev.* **82**, p. 538.

Slater, J. C. (1953). Electronic structure of solids. 1. The energy band method, Quarterly Progress Reports 4, Massachusetts Institute of Technology, Cambridge.

Smit, J. and Wijn, H. P. J. (1959). *Ferrites* (John Wiley & Sons, New York).

Solovyev, I. V. and Terakura, K. (2003). Orbital degeneracy and magnetism of perovskite manganese oxides, in *Electronic Structure and Magnetism of Complex Materials* (Springer, Berlin), p. 253.

Soulen, R., Byers, J., Osofsky, M., Nadgorny, B., Ambrose, T., Cheng, S., Broussard, P., Tanaka, C., Nowak, J., Moodera, J., Barry, A. and Coey, J. (1998). Measuring the spin polarization of a metal with a superconducting point contact, *Science* **282**, p. 85.

Souza, S. D., Saxena, R., Shreiner, W. and Zawislak, F. (1987). Magnetic hyperfine fields in Heusler alloys Co_2YZ (Y=Ti,Zr; Z=Al,Ga,Sn), *Hyperfine Interactions* **34**, p. 431.

Spinu, L., Srikanth, H., Gupta, A., Li, X. W. and Xiao, G. (2000). Probing magnetic anisotropy effects in epitaxial CrO_2 thin films, *Phys. Rev. B* **62**, p. 8931.

Stoffel, A. M. (1969). Magnetic and magneto-optic properties of FeRh and CrO_2, *J. Appl. Phys.* **40**, p. 1238.

Stöhr, J. (1999). Exploring the microscopic origin of magnetic anisotropies with X-ray magnetic circular dichroism (XMCD) spectroscopy, *J. Magn. Magn. Mater.* **200**, p. 470.

Stoner, E. C. (1939). Collective electron ferromagnetism. ii. energy and specific heat, *Proc. R. Soc. Lond. A* **169**, p. 339.

Stranski, I. N. and Krastanow, L. V. (1939). Abhandlungen der Mathematisch-Naturwissenschaftlichen Klasse, *Akademie der Wissenschaften und der Literatur in Mainz* **146**, p. 797.

Strijkers, G. J., Ji, Y., Yang, F. Y., Chien, C. L. and Byers, J. M. (2001). Andreev reflections at metal/superconductor point contacts: Measurement and analysis, *Phys. Rev. B* **63**, p. 104510.

Suzuki, T. and Ido, H. (1993). Magnetic-nonmagnetic transition in CrAs and the related compounds, *J. Appl. Phys.* **73**, p. 5686.

Tersoff, J. and Hamann, D. (1985). Theory of the scanning tunneling microscope, *Phys. Rev. B* **31**, p. 805.

Thamer, B. J., Douglass, R. M. and Staritzky, E. (1957). The thermal decomposition of aqueous chromic acid and some properties of the resulting solid phases, *J. Am. Chem. Soc.* **79**, p. 547.

Thomas, A. (2003). *Preparation and Characterisation of Magnetic Single and Double Barrier Junctions*, Ph.D. thesis, Bielefeld University, Bielefeld, Germany.

Tobola, J., Pierre, J., Kaprzyk, S., Skolozdra, R. V. and Kouacou, M. A. (1998). Crossover from semiconductor to magnetic metal in semi-Heusler phases as a function of valence electron concentration, *J. Phys.: Condens. Mat.* **10**, p. 1013.

Toney, M. and Brennan, S. (1989). Measurements of carbon thin films using x-ray reflectivity, *J. Appl. Phys.* **66**, p. 1861.

Turban, P., Andrieu, S., Kierren, B., Snoeck, E., Teodorescu, C. and Traverse, A. (2002). Growth and characterization of single crystalline NiMnSb thin films and epitaxial NiMnSb/MgO/NiMnSb(001) trilayers, *Phys. Rev. B* **65**, p. 134417.

Uhrig, M., Beck, A., Goeke, J., Eschen, F., Sohn, M., Hanne, G. F., Jost, K. and Kessler, J. (1989). Calibration of a Mott detector using circularly polarized impact radiation from helium, *Rev. Sci. Instrum.* **60**, p. 872.

van der Heidet, P. A. M., Baeldet, W., de Groot, R. A., de Vrooment, A. R., van Engent, P. G. and Buschow, K. H. J. (1985). Optical properties of some half-metallic ferromagnets, *J. Phys. F: Met. Phys.* **15**, p. L75.

van der Zaag, P. J., Bloemen, P. J. H., Gaines, J. M., Wolf, R. M., van der Heijden, P. A. A., van de Veerdonk, R. J. M. and de Jonge, W. J. M. (2000). On the construction of an Fe_3O_4-based all-oxide spin valve, *J. Magn. Magn. Mater.* **211**, p. 301.

van Dijken, S., Fain, X., Watts, S. M., Nakajima, K. and Coey, J. M. D. (2004). Magnetoresistance of $Fe_3O_4/Au/Fe_3O_4$ and $Fe_3O_4/Au/Fe$ spin-valve structures, *J. Magn. Magn. Mater.* **280**, p. 322.

van Engen, P., Buschow, K., Jongebreur, R. and Erman, M. (1983). PtMnSb, a material with very high magneto-optical Kerr effect, *Appl. Phys. Lett.* **42**, p. 202.

Van Roy, W., De Boeck, J., Brijs, B. and Borghs, G. (2000). Epitaxial NiMnSb films on GaAs(001), *Appl. Phys. Lett.* **77**, p. 4190.

Vanderbilt, D. (1990). Soft self-consistent pseudopotentials in a generalized eigenvalue formalism, *Phys. Rev. B* **41**, p. 7892.

Versluijs, J. J. and Coey, J. M. D. (2001). Magnetotransport properties of Fe_3O_4 nanocontacts, *J. Magn. Magn. Mater.* **226–230**, p. 688.

Verwey, E. J. W. (1939). Electronic conduction of magnetite (Fe_3O_4) and its transition point at low temperatures, *Nature* **144**, p. 327.

von Barth, J., Fecher, G. H., Balke, B., Ouardi, S., Graf, T., Felser, C., Shkabko, A., Weidenkaff, A., Klaer, P., Elmers, H. J., Yoshikawa, H., Ueda, S. and Kobayashi, K. (2010). Itinerant half-metallic ferromagnets Co_2TiZ (Z=Si, Ge, Sn): *Ab initio* calculations and measurement of the electronic structure and transport properties, *Phys. Rev. B* **81**, p. 064404.

von Barth, U. and Hedin, L. (1972). A local exchange-correlation potential for the spin polarized case. I, *J. Phys. C: Solid Stat. Phys.* **5**, p. 1629.

von Helmolt, R., Wecker, J., Holzapfel, B., Schultz, L. and Samwer, K. (1993). Giant negative magnetoresistance in perovskitelike $La_{2/3}Ba_{1/3}MnO_x$ ferromagnetic films, *Phys. Rev. Lett.* **71**, p. 2331.

Wang, F. and Vaedeny, Z. V. (2009). Organic spin valves: the first organic spintronics devices, *J. Mater. Chem.* **19**, p. 1685.

Wang, H.-Y. and Qian, M. C. (2006). Electronic and magnetic properties of Mn/Ge digital ferromagnetic heterostructures: An *ab initio* investigation, *J. Appl. Phys.* **99**, p. 08D705.

Wang, W. H., Przybylski, M., Kuch, W., Chelaru, L. I., Wang, J., Lu, Y. F., Barthel, J. and Kirschner, J. (2005a). Spin polarization of single-crystalline Co_2MnSi films grown by PLD on GaAs(001), *J. Magn. Magn. Mater.* **286**, p. 336.

Wang, W. H., Przybylski, M., Kuch, W., Chelaru, L. I., Wang, J., Lu, Y. F., Barthel, J., Meyerheim, H. L. and Kirschner, J. (2005b). Magnetic properties and spin polarization of Co_2MnSi Heusler alloy thin films epitaxially grown on GaAs(001), *Phys. Rev. B* **71**, p. 144416.

Wang, X., Antropov, V. P. and Harmon, B. N. (1994). First principles study of magneto-optical properties of half-metallic Heusler alloys: NiMnSb and PtMnSb, *IEEE Trans. Magn.* **30**, p. 4458.

Webster, P. J. (1971). Magnetic and chemical order in Heusler alloys containing cobalt and manganese, *J. Phys. Chem. Solids* **32**, p. 1221.

Webster, P. J. and Ziebeck, K. R. A. (1988). *Landolt-Börnstein—Group III Condensed Matter, SpringerMaterials: The Landolt-Börnstein Database,* Vol. 19c (Springer-Verlag, Berlin), p. 75.

Wei, J. Y. T., Yeh, N.-C. and Vasquez, R. P. (1997). Tunneling evidence of half-metallic ferromagnetism in $La_{0.7}Ca_{0.3}MnO_3$, *Phys. Rev. Lett.* **79**, p. 5150.

Westerholt, K., Bergmann, A., Grabis, J., Nefedov, A. and Zabel, H. (2005). Half-metallic alloys—fundamentals and applications, in I. Galanakis and P. H. Dederichs (eds.), *Lecture Notes in Physics* (Springer-Verlag, Berlin), p. 67.

Williams, A., Kübler, J. and Gelatt, C. (1979). Cohesive properties of metallic compounds: Augmented-spherical-wave calculations, *Phys. Rev. B* **19**, p. 6094.

Wojcik, M., Van Roy, W., Jedryka, E., Nadolski, S., Borghs, G. and De Boeck, J. (2002). NMR evidence for MnSb environments within epitaxial NiMnSb films grown on GaAs(001), *J. Magn. Magn. Mater.* **240**, p. 414.

Wolf, S. A., Awschalom, D. D., Buhrman, R. A., Daughton, J. M., Molnár, S., Roukes, M., Chtchelkanova, A. and Treger, D. (2001). Spintronics: A spin-based electronics vision for the future, *Science* **294**, p. 1488.

Wright, J. P., Attfield, J. P. and Radaelli, P. G. (2001). Long range charge ordering in magnetite below the Verwey transition, *Phys. Rev. Lett.* **87**, p. 266401.

Wright, J. P., Attfield, J. P. and Radaelli, P. G. (2002). Charge ordered structure of magnetite Fe_3O_4 below the Verwey transition, *Phys. Rev. B* **66**, p. 214422.

Wu, H., Kratzer, P. and Scheffler, M. (2005). First-principles study of thin magnetic transition-metal silicide films on Si(001), *Phys. Rev. B* **72**, p. 144425.

Wu, H., Kratzer, P. and Scheffler, M. (2007). Density-functional theory study of half-metallic heterostructures: Interstitial Mn in Si, *Phys. Rev. Lett.* **98**, p. 117202.

Wurmehl, S., Fecher, G. H., Kandpal, H., Ksenofontov, V., Felser, C., Lin, H.-J. and Morais, J. (2005). Geometric, electronic, and magnetic structure of Co_2FeSi: Curie temperature and magnetic moment measurements and calculations, *Phys. Rev. B* **72**, p. 184434.

Wurmehl, S., Fecher, G. H., Kandpal, H., Ksenofontov, V., Felser, C. and Lin, H.-J. (2006a). Investigation of Co_2FeSi: The Heusler compound with highest Curie temperature and magnetic moment, *Appl. Phys. Lett.* **88**, p. 032503.

Wurmehl, S., Fecher, G. H., Kroth, K., Kronast, F., Dürr, H. A., Takeda, Y., Saitoh, Y., Kobayashi, K., Lin, H.-J., Schönhense, G. and Felser, C. (2006b). Electronic structure and spectroscopy of the quaternary Heusler alloy $Co_2Cr_{1-x}Fe_xAl$, *J. Phys. D: Appl. Phys.* **39**, p. 803.

Wurmehl, S., Kohlhepp, J. T., Swagten, H. J. M., Koopmans, B., Wojcik, M., Balke, B., Blum, C. G. F., Ksenofontov, V., Fecher, G. H. and Felser, C. (2007). Probing the random distribution of half-metallic $Co_2Mn_{1-x}Fe_xSi$ Heusler alloys, *Appl. Phys. Lett.* **91**, p. 052506.

Wyckoff, R. W. G. (1963). *Crystal Structures, 2nd edition*, Vol. 1 (Interscience, New York).

Xie, W.-H., Xu, Y.-Q., Liu, B.-G. and Pettifor, D. G. (2003). Half-metallic ferromagnetism and structural stability of zincblende phases of the transition-metal chalcogenides, *Phys. Rev. Lett.* **91**, p. 037204.

Xu, Y.-Q., Liu, B.-G. and Pettifor, D. G. (2002). Half-metallic ferromagnetism of MnBi in the zinc-blende structure, *Phys. Rev. B* **66**, p. 184435.

Yablonskikh, M. V., Grebennikov, V. I., Yarmoshenko, Y. M., Kurmaev, E. Z., Butorin, S. M., Duda, L.-C., Sthe, C., Käämbre, T., Magnuson, M. and Nordgren, J. (2000). Magnetic circular dichroism in X-ray fluorescence of Heusler alloys at threshold excitation, *Solid State Commun.* **117**, p. 79.

Yablonskikh, M. V., Yarmoshenko, Y. M., Grebennikov, V. I., Kurmaev, E. Z., Butorin, S. M., Duda, L.-C., Nordgren, J., Plogmann, S. and Neumann, M. (2001). Origin of magnetic circular dichroism in soft x-ray fluorescence of Heusler alloys at threshold excitation, *Phys. Rev. B* **63**, p. 235117.

Yamasaki, A., Imada, S., Suga, S., Kanomata, T. and Ishida, S. (2002). Magnetic circular dichroism in the soft x-ray absorption spectra of Co-based Heusler alloys, *Surf. Rev. Lett.* **9**, p. 955.

Yamashita, J. and Kondo, J. (1958). Superexchange interaction, *Phys. Rev.* **109**, p. 730.

Yanase, A. and Siratori, K. (1984). Band structure in the high temperature phase of Fe_3O_4, *J. Phys. Soc. Jpn.* **53**, p. 312.

Yao, Y., Kleinman, L., MacDonald, A., Sinova, J., Jungwirth, T., Wang, D.-S., Wang, E. and Niu, Q. (2004). First principles calculation of anomalous Hall conductivity in ferromagnetic bcc Fe, *Phys. Rev. Lett.* **92**, p. 037204.

Ye, J., Kim, Y., Millis, A., Shraiman, B., Majumdar, P. and Tešanović, Z. (1999). Berry phase theory of the anomalous hall effect: Application to colossal magnetoresistance manganites, *Phys. Rev. Lett.* **83**, p. 3737.

Yoneda, Y. (1963). Anomalous surface reflection of X rays, *Phys. Rev.* **131**, p. 2010.

Yoshimura, K., Miyazaki, A., Vijayaraghavan, R. and Nakamura, Y. (1985). Hyperfine field of the Co_2YZ Heusler alloy (Y = V, Cr, Mn and Fe; Z = Al and Ga), *J. Magn. Magn. Mater.* **53**, p. 189.

Youn, S. and Min, B. (1995). Effects of the spin–orbit interaction in Heusler compounds: Electronic structures and Fermi surfaces of NiMnSb and PtMnSb, *Phys. Rev. B* **51**, p. 10436.

Young, R. A., Sakthivel, A., Moss, T. S. and Paiva-Santos, C. O. (1995). *DBWS–9411* – an upgrade of the *DBWS*.** programs for Rietveld refinement with PC and mainframe computers, *J. Appl. Cryst.* **28**, p. 366.

Zeller, R. (2006). Spin-polarized DFT calculations and magnetism, *Computational Nanoscience: Do It Yourself!* **31**, p. 419.

Zener, C. (1951). Interaction between the *d*-shells in the transition metals. II. ferromagnetic compounds of manganese with perovskite structure, *Phys. Rev.* **82**, p. 403.

Zhang, F. M., Liu, X. C., Gao, J., Wu, X. S., Du, Y. W., Zhu, H., Xiao, J. Q. and Chen, P. (2004). Investigation on the magnetic and electrical properties of crystalline $Mn_{0.05}Si_{0.95}$ films, *Appl. Phys. Lett.* **85**, p. 786.

Zhang, W., Jiko, N., Mibu, K. and Yoshimura, K. (2005). Effect of substitution of Mn with Fe or Cr in Heusler alloy of Co_2MnSn, *J. Phys.: Condens. Mat.* **17**, p. 6653.

Zhang, Z. and Satpathy, S. (1991). Electron states, magnetism, and the Verwey transition in magnetite, *Phys. Rev. B* **44**, p. 13319.

Zhao, J. H., Matsukura, F., Takamura, K., Abe, E., Chiba, D. and Ohno, H. (2001). Room-temperature ferromagnetism in zincblende CrSb grown by molecular-beam epitaxy, *Appl. Phys. Lett.* **79**, p. 2776.

Zhao, Y.-J., Geng, W. T., Freeman, A. J. and Delley, B. (2002). Structural, electronic, and magnetic properties of $\alpha-$ and $\beta-$MnAs: LDA and GGA investigations, *Phys. Rev. B* **65**, p. 113202.

Zhu, W., Sinkovic, B., Vescovo, E., Tanaka, C. and Moodera, J. S. (2001). Spin-resolved density of states at the surface of NiMnSb, *Phys. Rev. B* **64**, p. 060403.

Zhu, W., Zhang, Z. and Kaxiras, E. (2008). Dopant-assisted concentration enhancement of substitutional Mn in Si and Ge, *Phys. Rev. Lett.* **100**, p. 027205.

Ziebeck, K. R. A. and Webster, P. J. (1974). A neutron diffraction and magnetization study of Heusler alloys containing Co and Zr, Hf, V or Nb, *J. Phys. Chem. Solids* **35**, p. 1.

Ziese, M. and Blythe, H. J. (2000). Magnetoresistance of magnetite, *J. Phys.: Condens. Mat.* **12**, p. 13.

Index